There Is No Theory of Everything

Lars Q. English

There Is No Theory of Everything

A Physics Perspective on Emergence

 Springer

Lars Q. English
Department of Physics & Astronomy
Dickinson College
Carlisle, PA, USA

ISBN 978-3-319-86556-0   ISBN 978-3-319-59150-6   (eBook)
DOI 10.1007/978-3-319-59150-6

Printed on acid-free paper

This Springer imprint is published by Springer Nature
The registered company is Springer International Publishing AG
The registered company address is: Gewerbestrasse 11, 6330 Cham, Switzerland

Preface

The last time I visited the Museum of Modern Art in New York, something caught my eye that had escaped my notice on a previous visit. Perhaps I happened to be in the right frame of mind that day to receive the message, or maybe I was just following the coordinated motion of the throngs of visitors. As the case may be, I stopped, and there in front of me on a large white canvas were printed the following words in capitalized black letters:

DO YOU SENSE HOW ALL THE PARTS OF A GOOD PICTURE ARE INVOLVED WITH EACH OTHER, NOT JUST PLACED SIDE BY SIDE? ART IS CREATION FOR THE EYE AND CAN ONLY BE HINTED AT WITH WORDS

The work was by American artist John Baldessari and entitled "What is Painting." I read it again. There was something about that piece that definitely commanded attention. What specifically was it, though, that intrigued me, that drew me in? Could it have been the oblique self-reference which, on second thought, I realized was actually quite clever, even witty? Or perhaps its appeal was more philosophical in nature. Did it resonate with a larger question I had been pondering lately? Did it speak to the ways in which "meaning" could emerge from the grouping of things that taken by themselves seemed to lack it?

In my first week of college, I had encountered a dramatic example of "meaning" suddenly arising from a collection of random things. As I remember it, my math professor had handed out sheets of paper with small dots littering the page. When I looked at it, no intelligible patterns could be discerned whatsoever, no matter how hard one tried. There were just thousands of dots seemingly randomly strewn on the page. I was about to dismiss the whole thing as a silly stunt by a slightly eccentric professor when he asked us to hold up the handout in front of us and focus our eyes on a point behind the page. It was actually not easy to do—my eyes just naturally wanted to focus on the dots themselves. On some level, they must have still been seeking some subtle pattern in their midst. I tried but still couldn't quite let go of the dots, visually. I heard isolated shouts of "wow" in the classroom. Some people were clearly getting something I was still utterly missing. I forced myself again to override my instinctive autofocus, and suddenly, there I saw it. A switch had been flipped, and now—as if by magic—a three-dimensional image leaped out from the page. From the seemingly random collection of dots, a detailed image had instantaneously formed in front of me.

Since that time, the quality of such images has continuously improved (color was added along the way), and they have become rather commonplace, but the effect still manages to surprise and startle a bit. And the basic question remains: Where does the "meaning" reside? Does its representation reduce down to the individual dots on the page? Is it in the individual dots, the dots as a whole, or the relationship between the dots? How does the picture get constructed from the individual dots? These are questions not unlike ones explored in contemporary science as well, although the word "dot" is usually replaced with "atom," or "particle," or "cell." Matter is, of course, made up of elementary particles, but it does seem to acquire properties that are not easily traced back to them. Rather, it is often the way in which these particles "are involved with each other" that counts. Our bodies are made up of cells, but they acquire abilities that far transcend the individual cells and arise as a result of highly complex organization across many scales.

The observation that "the whole is greater than the sum of its parts" and that "new things happen when parts interact to form systems" generally goes by the name of *emergence*. One aim of this book is to illustrate how ideas and discoveries in various branches of modern physics have informed and sharpened the ways we think about emergence. Along our conversational tour through the terrain of quantum mechanics, condensed-matter physics, and nonlinear and statistical physics, I hope to highlight the various nuances that arise in our encounters with emergence. I also hope to convey a sense that the layers added by various physical theories give depth, contrast, and definition to this framework of emergence. Such established framework allows us to then reflect on some larger lessons that emergence affords us.

One of those larger lessons is the realization that the great diversity of theories and models, the myriad of phenomena and explanatory approaches, will always be with us in the sciences, as well as in areas beyond science. There is no final "theory of everything" just waiting to be discovered. No theory, however potent, could in one fell swoop explain away the complexities that arise across different scales or predict the nuances that present themselves in various observational contexts. The idea that such a theory may be just around the corner is incompatible with notions of emergence that science itself has uncovered. The organization of science in terms of disciplines and departments will undoubtedly change over time, but we can be pretty confident that at no time will string theory be the only thing we have to teach science students. Atomic physics cannot yield insight into the laws of macroeconomics, nor does cell biology help us sort out high-temperature superconductivity in any way. Scientific context matters a great deal, and phenomena generally resist overly reductive explanatory approaches. This view also challenges notions of a scientific hierarchy flowing from fundamental to applied—emergence is the great equalizer among science disciplines.

While the book explores notions of emergence within the physical sciences, and hopes to showcase their intellectual power, it is not meant to be antireductionist. It does not claim that reductionism, in the sense of striving for general, broadly applicable principles or exploring the microcosm, is flawed. Without a doubt, enormous progress has been made along these lines in physics. Instead, I want to examine both practical and foundational limitations of the reductionist "divide-and-conquer" strategy within many areas of contemporary science and to challenge the often unquestioned assumption that only within ever smaller constituent parts can we hope to discover truly novel science. But this should not be interpreted as an endorsement of the strongest notions of ontological emergence, or of strong *emergentism*, which (as I argue) tend to overshoot the mark.

After a broad overview in Chap. 1, we begin at the small scale in Chap. 2 with a look at the quantum view on particles, entanglement, and its implications for emergence. Then, in Chap. 3, we take a look at what happens when atoms aggregate to form the macroscopic matter we are familiar with, as well as more exotic states of matter at lower temperatures. Here, we will encounter examples that emphasize the importance of structure and organization over mere composition and that illustrate feedback from large to small. Chapter 4 then follows up with a discussion on modern approaches to phase transitions which arguably put emergence on a firmer theoretical foundation; although the discussion is colloquial, here we probably come to the most technical part of the book.

We then shift gears in Chap. 5 and look at nonlinear systems, chaos theory, and spontaneous pattern formation. In this field, we encounter two quintessential phenomena that convey different aspects of emergence—animal swarming and convective motion of fluids. The former shows how the local adherence to simple rules gives rise to remarkably intricate global dynamics, whereas the latter is an example of a stronger form of downward control. Chapter 6 then examines the role of effective models in science that work well within a certain context and scale, but get supplanted by different approaches outside of them. The neuron will provide us with a good example illustrating this point.

We will briefly look at sociology in Chap. 7—the first field of study to formulate explicit notions of emergence—and examine the importance of social networks, norms, and their downward action. One of the broader insights here is that nonreductive reasoning in this arena can provide powerful antiracist correctives to certain pernicious social narratives. In Chap. 8, we will seek to apply lessons of emergence to another, perhaps more remote, place—the relationship between science and religion. Emergence would imply here that religion cannot be reduced to psychology, which in turn is not reducible to neuroscience and so on. We will argue that neither is religious experience fully accessible to scientific analysis nor will there be a grand unification of the two, as envisioned by natural theology, for instance.

A larger lesson of emergence that weaves through much of the book is the proposition that there can be multiple explanatory frameworks occupying different dimensions of human inquiry and understanding, each discovering *emergent truths* within their own respective domains. Even when different frameworks meet, the important interdisciplinary work at the boundaries of disciplines does not fully merge their descriptions into one.

Carlisle, PA, USA   Lars Q. English

Acknowledgments

I would like to acknowledge the following people. Beverly West (Cornell University) was very kind in reading a preliminary version of the full manuscript, providing numerous and very valuable feedback. Ron Winters (Denison University) graciously allowed me to engage him in extended conversations during and immediately following my visit to Denison. He also gave me plenty of pointers and encouragement on an early draft of Chaps. 2, 3, and 4. I thank my colleague at Dickinson College, Jeff Forrester, for his constant stream of insightful ideas and intriguing connections, some of which made their way into Chap. 6. Last, but certainly not least, I want to thank my wife, Anita Mareno (Penn State), who read the book in its various stages, always pointing out possible ways for stylistic and substantive improvement. I am also grateful for her unwavering patience and encouragement. Finally, I acknowledge the support by my home institution, Dickinson College, and the hospitality of SUNY New Paltz, where a part of the book was written.

Contents

Chapter 1
First Encounters

1.1 Physical Vantage Points

The experimental discovery in 2012 of the Higgs boson—previously the last unconfirmed particle postulated by the Standard Model of particle physics—was greeted with huge relief and lavish celebration among physicists around the globe. An international search party—conducted both at Fermi Lab and at CERN—had repeatedly come up empty-handed over the span of decades, and now finally the elusive particle had been spotted by two enormous detectors within the inner sanctum of the Large Hadron Collider at CERN. The fear that physicists might have been chasing a phantom all those years proved unfounded. No— the vast army of researchers and technicians had, in fact, been steadily closing in on a most-wanted particle.

Not only was the physics community understandably happy, the particle's confirmation was also widely reported in the press and seemed to resonate with the larger public. It didn't hurt that there was a compelling human-interest story to be told. Peter Higgs had postulated the existence of the particle that bears his name back in 1964, but was then held in suspended animation for nearly five decades until the ripe age of 84, when finally the good news arrived. Moreover, "the God particle", as it became affectionately known to most folks, seemed to arouse broad curiosity and stir the collective imagination due to its

© Springer International Publishing AG 2017
L.Q. English, *There Is No Theory of Everything*,
DOI 10.1007/978-3-319-59150-6_1

almost mythical-sounding powers. Here was a particle that was said to bestow mass on all others, and verification of its existence seemed to imply that perhaps the fundamental origin of mass had finally been uncovered.

What these experiments at CERN did not do is to discover anything beyond the Standard Model. Even at the staggering collision energies reached in the 27 km-long tunnels of the LHC, every observation to date has been in line with its theoretical prediction. That fact alone is a testament to the enormous intellectual achievement that is the Standard Model. More broadly, it perfectly exemplifies one of the big drivers of progress in physics, namely the quest to discover fundamental laws and properties.

In fact, the "big" breakthroughs in theoretical physics have all involved, as Richard Feynman put it, *amalgamations* [1], where phenomena that had appeared as disparate or unrelated were now shown to be simply different sides of the same coin, different manifestations of the same underlying principle. Certainly, this was true in the case of James Clerk Maxwell's electromagnetic theory and light in the late nineteenth century, which unified electricity and magnetism, and showed that visible light was an electromagnetic wave and that its color was simply a measure of the wavelength of this electromagnetic wave. Another significant amalgamation came with the unification of thermodynamics and statistical mechanics around the same time, and with Einstein's special theory of relativity of space-time. More recently, physicists have been successful in unifying at first two of the four fundamental forces of nature (creating the electroweak force), then adding the strong force into the mix as well. Currently, efforts within String Theory are under way to formulate a quantum theory of gravity which so far has proved elusive but would perhaps lead to a full unification of all four forces.

Unity is aesthetically pleasing to scientists, and perhaps especially so to physicists: when the world with its myriad of phenomena, its "messiness", bewildering complexity and unending variety, can be shown to obey just a few principles at the core, it feels intellectually gratifying. In my own discipline of physics, if I had to pick some of the overarching themes that tie things together across all areas of specialization, I would probably point to ideas like symmetry (and symmetry-breaking), its connection to conservation laws, reversibility and irreversibility of processes, the notion of the *field*, and so on. These ideas organize our knowledge and give it coherence and structure. They constrain what is possible, and guide the models we construct of phenomena, at the same time allowing us also to systematically dismiss other explanations that don't square with those ideas. There can be no doubt that the reliance on well established guiding principles is essential to the scientific enterprise. If that were the only definition of reductionism, we would all be avid reductionists in science.

But reductionism, in the published views and assertions of its proponents and in the general public discourse, has taken on a much more targeted meaning. It has now come to be associated with the view that nothing truly new arises across higher levels of scale, organization or complexity, and that the intricate and rich behavior exhibited by larger systems is completely traceable to, and can be understood via, the properties of the constituent parts and the rules governing them in isolation. In physics, this view gets expressed most often by people bemoaning the end of our discipline after a "final theory of everything" will have been formulated; what else could there possibly be left to do [2]? In neuroscience, we see it in the unquestioned assumption that in order to understand high-level cognitive functions, or even consciousness, we must study the brain at the level of the single neuron (or interneuron communication); that whatever the brain is capable of must somehow already be contained there [3]. It is considered insufficient or incomplete to discover the rules governing complex systems through careful experimentation if these rules cannot be deduced or derived directly from a more fundamental set of laws operating at a more basic level.

An alternative narrative has increasingly taken hold in the physical sciences. It is a philosophical framework, supported by large swaths of modern science, that goes by the name of emergence. Emergence opposes the reductionist premise of an unbroken path from 'applied' to 'fundamental', asserting instead that non-reductive behavior arises at different levels of organization and complexity, and within different scientific contexts. The behavior of complex systems, in turn, is described by scientific rules and principles which cannot be derived or predicted from the science governing the system's parts in isolation. The next chapters will attempt to elucidate the notion of emergence and its implications by examining how it manifests in various scientific contexts, such as in quantum and condensed-matter physics and nonlinear science, but one of its central contentions can already be stated. That is, there can be no theory of everything. No single theory could explain, however indirectly or remotely, the myriad of phenomena we observe across the spectrum of human inquiry.

In contrast, the reductionist quest is all about uncovering an imagined "theory of everything." Here, science is believed to find its highest and purest expression in the discovery of ever-more "fundamental" laws, considered the milemarkers along the road to such a final theory. Perhaps not surprisingly, this journey has led deeper and deeper into the microcosm. After all, where else would the most fundamental laws be hiding? Where else could we expect to find this theory of everything? The things within our macroscopic world are made up of an unimaginably large number of atoms. A piece of graphite weighing only $12\,g$ is made up of about 6×10^{23} carbon atoms. That's a 6 with 23 zeros added to it. It is not farfetched to think that the "messiness" of things in our world comes from the great variety in which the basic building blocks can be put together. Thus, in order to get to anything fundamental, we have to study those building blocks and their interactions. If we understood nature at that level, then according to the reductionist program we could construct the rest of the world from the bottom up. In this way, we feel as though we could finally grasp and comprehend the entire world around us through these simple fundamental rules alone. This promise holds some seductive appeal for many of us in science, and we begin to succumb to the expectation that the world should be comprehensible solely through the lens of basic principles.

Granted, science has no doubt gained a tremendous amount of knowledge following this instinct. By studying atoms in detail, quantum theory was developed in the first part of the twentieth century. We now understand atoms as made up of a small, compact, positively charged nucleus around which are the negatively charged electrons. These electrons can be in discrete energy states all of which can be visualized as clouds around the nucleus with predictable shapes. Using this theory of electrons in atoms, we apply it to bonding between atoms, which is due to the overlap of electron clouds between nuclei, and we find ourselves entering the field of chemistry. Or we could keep going down in spatial scale and investigate the nucleus itself. There we learn that the nucleus is made up of protons and neutrons, and that these themselves are made up of even smaller particles called *quarks*. Now we are squarely in the field of elementary particle (also called high-energy) physics, where we learn furthermore that these quarks interact via an exchange of "messenger particles"—aptly called *gluons*—to produce the strong force.

This is all well and good, but it is worth remembering that chemistry flourished as a science well before the precise mechanism for the strong force was understood, and even before the strong force was even conceived as a fundamental force of nature. So while elementary-particle physics has opened up fascinating new worlds at the very small scale, much of its discoveries have proved irrelevant to the laws and processes of chemistry. Chemistry, as a discipline, has not had to adjust in the face of new and important results from particles and fields. The same is also true for other branches of physics. Nothing within condensed-matter physics had to be revised or reinterpreted after the electroweak unification had succeeded, or after the Higgs boson was found. It is fair to say that elementary particle physics has not had the kind of broad impact on all of science that we might naively expect. This is our first glimpse that while reductionism may hold some aesthetic appeal for us, in the real world it may not operate exactly as advertised; the practical reality is different and quite a bit more nuanced.

In 1972 an article appeared in the journal *Science* with the title "More is Different" [4] written by the physicist Philip Anderson. Anderson was at that time already one of the most highly regarded theorists in physics credited with influential work in many areas of fundamental material science, including contributions to superconductivity, electrons in disordered materials, low-temperature properties of glasses, and magnetism. What had topped it off was the ability to transfer his deep knowledge of condensed-matter physics into the context of elementary particles.

To put things in a historical context, the age of the all-rounded physicist had really mostly ended at that point. Physics as a discipline had advanced so rapidly that by the 1970s it was quite rare for anyone—even the most celebrated of individuals—to make significant contributions in two or more broad sub-fields of physics. Hans Bethe had been such a "renaissance man" of physics. He is perhaps best known for his work in nuclear physics and astrophysics, explaining for the first time what makes stars shine (nuclear fusion) in the 1930s, but his name is equally associated with many concepts in solid-state physics, as well as for his early contributions to quantum electrodynamics. In retrospect, Bethe was really part of a dying breed; he was perhaps one of the last of the famous polyglots of physics. The time of increasing specialization had begun. Against this backdrop, Philip Anderson's feat appears even more remarkable. His work suggested the possibility of *symmetry breaking* as a means for particles that seemed like they should be mass-less to acquire mass nonetheless. Armed with insights from condensed-matter physics, this paper gave the initial impetus for what we now call the Higgs mechanism.

To make a long story short, when the now-famous article appeared in 1972, in which he set out to formulate for the fist time the scientific case for a version of emergence and the limits of reductionism, people were ready to listen:

> The ability to reduce everything to simple fundamental laws does not imply the ability to start from those laws and reconstruct the universe. In fact, the more the elementary particle physicists tell us about fundamental laws, the less relevance they seem to have to the very real problems of the rest of science, much less to those of society.

Why is it practically impossible to "reconstruct" the world from the fundamental laws of nature? Why do most phenomena and the conceptual structures to explain them seem emergent? Anderson used the important notion of *symmetry breaking* to illustrate this difficulty. In a nutshell, symmetry breaking refers to the well-known fact that larger-scale systems usually exhibit states of lower symmetry than what is implied by the fundamental laws that govern them. This does not, however, imply that larger-scale phenomena are in contradiction with the lower-level laws. We will come back to this point later and look at symmetry and phase transitions more closely.

For now, let us simply illustrate the basic idea of symmetry breaking via an everyday example—traffic. Let's feign ignorance for a second and ask a silly question: Why are all the cars on North American roads always driving on the right side? The answer is really pretty obvious, but let's pretend we are aliens having just made this observation of striking regularity. Now we ask: Does it have to do with the cars themselves? Are they built such that they can physically only drive on the right side of the road? Curiosity getting the better of us, we decide to take one of those cars apart, and we quickly come to the conclusion that no, they could equally well drive on the other side. In fact, at the level of the rules governing these cars and their driving mechanics, we find only symmetry: both sides are equally possible. (If it bothers you that there is a "driver's side" with cars, then consider motorcycles instead.) Nevertheless, in real life that symmetry seems to be broken because we only see right-side driving.

The culprit is, of course, organizing principles internalized by drivers and enforced by law. But there is still a hint of the underlying symmetry: other countries, such as Britain and Japan, drive on the left side. The point is that whether a country decides to drive on the right or left cannot be traced back to the fundamental laws of cars or driving. These laws embody only perfect symmetry. Something else beyond fundamental law explains the symmetry breaking that we observe. As we will explore throughout the book, that 'something else' is larger-scale organization and downward control.

Emergence is often summed up by the adage that "the whole is more than the sum of its parts," and indeed the relationship between parts and whole is essential in any discussion on the topic. So what is the nature of this relationship? To put it succinctly, the parts, of course, make up the whole, but the whole also influences the behavior of the parts. It is a two-way street of dependent relationship. A system is made up of constituent parts and these parts may be comprised of yet smaller parts. At every level, parts interact with one another, and it is the relational interaction of the parts more so than their intrinsic property that is important for the behavior of the system as a whole. Furthermore, the interaction between parts is shaped, constrained, or dictated to some extent by the system's architecture, its "organizational principle" [5], or its "relational structure" [6]. Structure, architecture, organization— those are notions that do not exist at the level of the individual part; they only exist at the level of the system itself. And yet, they seem to act back on the individual parts to affect their behavior.

That structure rather than composition can be much more important in a system's behavior and properties is well known to the organic chemist. Isomers are different geometric configurations of the same collections of atoms. This means that the molecular formula for two isomers of propanol and methoxyethane, for instance, are the same: C_3H_8O, but the structures are different (see Fig. 1.1).

It turns out, perhaps not surprisingly, that different isomers can function very differently. In the example above the first isomer represents an alcohol, whereas the second isomers, lacking the OH group, is very different and acts chemically like an ether. Amazingly, the difference between isomers can be as minor as chirality in order to have a noticeable effect. Chirality differentiates between left-handed or right-handed versions of the same molecule—two versions that can only

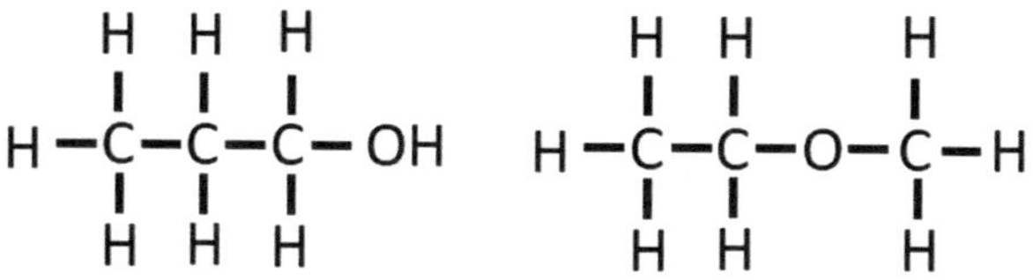

Fig. 1.1 Two different isomers containing the same set of atoms. The first arrangement is an alcohol (due to the OH group at the end), whereas the second is an ether. Chemically, the two molecules behave very differently

be made identical upon mirror inversion. Even this small difference, however, can have significant biological repercussions. As described vividly by Roald Hoffmann [7], our sense of taste is very much tied to chirality. Whereas a particular molecule may have taste, its mirror image is tasteless. Even the toxicity of a molecule can sometimes be tied to its chirality.

1.2   Here Come the Philosophers

As the isomer example illustrates, to claim that the properties of an object are solely determined by the characteristics of the parts of that object is clearly a fallacy. There are many examples closer to home that also confirm structure (rather than composition) to be a strong determining factor. We all know that people usually act differently at work than in a social setting. Employers have certain rules of engagement that they demand of their employees. You can't talk to your boss just like you would to a co-worker, and you would perhaps interact with an assistant assigned to you differently yet. Organizational structures usually rely on such hierarchies that then modify and regulate people's interactions with one another. These hierarchical structures don't exist inherently but only make sense in the context of the organization. If I decided to quit my job and leave the company, my boss would no longer be my boss.

So it appears that a higher-level phenomenon—the company—can modify the behavior at the lower level, the individual people that make up the company. If the very same individuals were not part of the entity called the company, they would relate to each other very differently. There is something about the company itself—it almost appears to possess causal powers on people. Think of the coworker of yours who was just promoted and given the title of Vice President. Chances are good that this person now relates to you very differently all of a sudden. The new position within the larger institution, the mere title, seems to drive or dictate a different inter-personal behavior. Gone is the slap-stick banter around the vending machine; instead you are now the recipient of dry emails outlining company policy and expectations.

Perhaps we are still a little puzzled. What exactly do we mean when we say a high-level phenomenon is emergent? What does emergence actually imply? Is it sufficient that we are practically prevented from predicting the high-level behavior from lower-level processes and laws due to intractable complexities? We might label this hypothesis weak emergence, or "epistemological emergence," because the limitation is only in our knowledge or cognitive ability, but it is not intrinsic or fundamental. Here one could argue that in reality low-level processes still do cause high-level behavior and phenomena, we just can't sort out exactly how. Upward causation from basic to composite, from micro to macro, still rules supreme, and emergent properties of complex systems are just an appearance, an illusion, an *epiphenomenon*, as philosophers would say. In this framework, reductionism is left intact at heart and deemed (however reluctantly) ultimately correct.

Alternatively, we could go further and claim that emergence implies an impossibility, even in principle, to go from the basic to the emergent level. This strong version of emergence has also been referred to "ontological emergence," and it usually relies on accepting high-level, emergent phenomena as possessing *causal powers*. If one accepted that a high-level phenomenon can cause another such phenomenon, or better yet, that it can cause something at a lower level—a notion termed *downward causality*—then the emergent phenomenon would exist on equal footing with a lower-level phenomenon. Neither one could be considered more fundamental, elementary or primary.

So, how seriously should we take notions like high-level causal power or downward causation? Empirically, it appears that downward causation is all around us, whether it be the workplace example from before, ecosystems creating selective pressures on species, or computer chips that would not exist were it not for the development of complex societal and technological structures [8], [9, Chap. 14]. The question debated in philosophical circles is how to make sense of downward causation on a deeper level [8–10]. Not surprisingly, the answer turns on what exactly we mean when we say A causes B.

One possible argument for a stronger form of emergence is to say that for predictive accuracy we have to adopt the view that the emergent phenomenon is a singular agent with causal power; it acts as one entity, even though it is made up of a multitude of parts. As a concrete (albeit improbable) example, let's say we wanted to predict what would happen if Canadian troops crossed into the U.S. uninvited. Would it make sense to analyze the situation from the point of view of the individual property owners whose land had been occupied by individual soldiers to see what their mutual interactions might be? Or would a more successful prediction acknowledge the reality of nations, their territorial claims and military capabilities. I think we can all agree that the latter approach would prove much more successful.

This approach to emergence has been called the "strategy of interpretation" [11]; we might also call it *functional emergence*. As a scientist, I am inclined to be satisfied with viewing emergence through this functional or operational lens. It is probably correct to say that in science, most of us are more concerned with finding out how things function and interact than we are with investigating their ultimate nature. Philosophers, however, do care about nature.

Granting that situations may be most fruitfully *interpreted* at the emergent level, one could still raise the objection that the high-level phenomenon might not possess independent reality, so to speak, with intrinsic properties [11]. But this is too high a bar to set, too extreme of an expectation. The whole will always rely on its parts for its existence, so in this sense the whole cannot exist "from its own side," as Buddhists would say. The whole always depends on the parts coming together, even if it modifies the behavior and character of these parts.

In short, the hypothesis of what has been referred to as "strong downward causation" [9, Chap. 1] may be somewhat problematic. Here the high-level phenomenon is seen to exist ontologically as a separate entity from its constituents, separate even from the laws governing them. This ontological separation, sometimes also called *reification* in philosophy, is not realistic. Without the parts and their interactions there is no whole, just as without the whole there are no parts (of the whole). Buddhists speak of the *emptiness* of singularity and plurality [12, pp. 123–128]. Nor can we posit that the whole can operate under rules that explicitly contradict the rules governing its parts. This is not something that scientists would accept. Philosophers refer to it as a violation of the *inclusivity* of levels [9, Chap. 1]. It's not as if the emergence of higher levels of complexity can directly invalidate the physical laws operating at the smaller scales.

The example often given is that of a biological cell which is, of course, made up of molecules, but which also produces macromolecules—proteins—that would otherwise not exist in nature. In other words, the actual material realization of the protein depended on the prior existence of the cell—a higher-level phenomenon. However, it is still true that these proteins do not violate any chemical laws, nor does the process by which the cell manufactured them. Everything happening within the cell is consistent with the microscopic laws of chemistry. The same is true for the examples from condensed-matter physics we will discuss later—superconductors are radically different from regular conductors, with respect to macroscopic behavior as well as microscopic organization, but they still obey the laws of quantum mechanics.

What we can perhaps say is that the biological cell allows certain microscopic configurations to manifest at certain times and not others; it conditions which microstates get realized and in what sequence. From the vast number of microstates potentially available to the system, the presence of complex structures may dictate which ones get actualized, and it may also govern how the next configuration at the micro-level arises from the previous one, guaranteeing temporal continuity.

Perhaps it is best to view the relationship between whole and parts, between basic and emergent phenomena, as depending on each other and arising together. Everyone has had the experience of being startled by the shrill noise that is produced when in a sound system the microphone gets too close to a speaker. Most people have also heard the term acoustic *feedback* used in this context. But what is really going on to produce this high-pitched noise? We can ask what would happen if the microphone picked up some small noise from somewhere. Following the sound-system's electronics, this noise would next be amplified and then send to the speaker. End of story? No, because the speaker is close to the mike, and so the mike will pick up whatever comes out of the speaker, including the amplified noise. So the noise gets amplified again, arrives at the speaker again, and gets picked up by the mike again, gets amplified yet again, and so on. Very quickly we reach saturation of the amplifier and we are left with a loud unpleasant noise.

So what started it all? Was it that the microphone picked up a stray sound, or was it that the speaker produced a stray sound at first? It doesn't actually matter which came first—the outcome is the same. The real culprit is sufficient feedback between the speaker and the microphone. The two phenomena of the speaker producing noise and the mike picking up the noise co-emerge to produce the shrill squeak.

We can think of emergence in a similar manner, only that the feedback now operates between whole and parts. This kind of feedback is reminiscent of Hofstadter's *strange loops* [13]. The parts aggregate to form the whole, but in the process of forming a larger system, they themselves are altered in some fashion. This change at the level of the parts, of course, immediately modifies the system properties again, which in turn acts back on the parts. Eventually, if a stable phenomenon is observed, an equilibrium establishes itself, and we can say that two phenomena—one at the low level and one at the high level—arose together. In this context, it helps to avoid the particular mental tendency of visualizing these processes as strictly ordered in time. Co-evolution is actually an apt descriptor, since we should not automatically think that first come the parts affecting the whole and then, later, comes the whole affecting the parts. Upward and downward causation should best be thought of as occurring together at the same time, or near the same time (with a miniscule time delays that are usually negligible). We will come back to the role of feedback and consistency between levels when we tackle phase transitions in Chap. 4.

Much has been made of the emergence of consciousness, and the notion that mind itself arises from matter (or neuronal activity and the like). Applying this idea of co-emergence to mind and matter, we might be led to the conclusion that mind emerges from matter while also acting back on its material substrate. More recently, philosophers have wrestled with foundational questions concerning this view and elucidated certain problems that come with it [8, 14–16]. It is certainly not an easy topic—neuroscientists and philosophers have been debating it for decades, and some religions weigh in as well. What is clear is that certain mental states have correlates in neuro-physiological brain activity, as can be demonstrated conclusively with functional MRI scans.

Beyond that, the nature of mind (or consciousness) and the type of downward control it exerts on the material substrate is seen differently across the spectrum of human inquiry. Neuro-scientists, of course, hold the ambition that one day consciousness will be understood from a scientific perspective. There is reason to remain skeptical: What seems to distinguish mental states is their subjective nature. Awareness is a quintessentially subjective experience, and as such it may be out of the reach of objectifiable scientific analysis. Of course, I do not mean to imply that the functioning of the brain cannot be investigated scientifically and understood in that way—much progress has already been made in that arena. But brain activity certainly is not identical with mind in the sense of subjective experience, of *qualia*. There is a qualitative difference of type between the two—something that a number of philosophers have pointed out recently [15, 16].

A consistent theme throughout the book is that an important lesson from emergence should be taken to heart: the essence of a phenomenon, what one might refer to as *meaning*, can only be captured at a commensurate level of description. While not denying that small-scale processes are involved in a phenomenon's appearance, its essential character cannot be located there, nor can its functional behavior be fully understood at that level. For all intents and purposes, autonomous worlds can arise across different spatial scales and levels of complexity, governed by effective laws that work well within their confines but lose meaning outside of them. Within science, as we will see, there are many such worlds already—biology is not reducible to string theory in physics, for instance. Emergence allows the biological sciences to carve out separate realities by effectively shielding them against encroachment by laws operating at smaller scales. This shielding, on the other hand, does not imply that there are no connections to related disciplines, nor does it in any way invalidate important work at their boundaries. But it does mean that no single theory with a unified language or description could, even in principle, account for all phenomena within physics and biology. There cannot be a final theory of everything.

1.3   A Brief History

If emergence is becoming increasingly accepted in broad swaths of the scientific community, that message has not been received more broadly. Most people do not seem aware that a paradigm shift is occurring. In larger society, science is still mostly seen as advocating for, and advancing along, reductionist lines. Why is that? It's an interesting question to ask. Why is reductionism still so popular, and why is it so frequently associated with the physical sciences?

One immediate answer might be that there have been many times in the past where the approach of reduction did turn out to be a successful strategy. Analyzing a system by isolating a very small number of its parts and subjecting them to careful study has been standard practice in science since its inception. It's a fact that small sub-systems more easily reveal the basic principles at play. Once those principles are understood, they can be fruitfully applied to the more complicated larger system. That's the basic idea that oftentimes works very well. Only more recently has science concerned itself with problems less amenable to this approach. Previously, however, physics largely limited itself to problems that could be addressed in this manner—problems that were susceptible to the 'divide and conquer' strategy. And that story begins (as so many in science) with Newtonian mechanics and its clockwork universe.

In the aftermath of the Newtonian revolution our ideas about the inner workings of the solar system and our place in the larger universe began to change, and this in turn sparked a huge burst of technological and innovative activity leading directly to the Industrial Revolution. James Watt's steam engine would not have been invented without the flourishing of science and engineering brought about by the monumental shift in paradigm of Newtonian mechanics.

But what the Newtonian revolution also brought with it was the notion of the universe as a deterministic clockwork; a clock that once set in motion would run true forever. A powerful version of the fully deterministic view was expressed by French mathematician and astronomer Pierre Simon Laplace in the 1820s and is often quoted as illustrating the extent to which determinism had begun to dominate early nineteenth century thinking:

> We may regard the present state of the universe as the effect of its past and the cause of its future. An intellect which at a certain moment would know all forces that set nature in motion, and all positions of all items of which nature is composed, if this intellect were also vast enough to submit these data to analysis, it would embrace in a single formula the movements of the greatest bodies of the universe and those of the tiniest atom; for such an intellect nothing would be uncertain and the future just like the past would be present before its eyes. [17]

In today's context, of course, such strident expressions of determinism strike most of us as naive, at best, and would be dismissed as untenable by scientists, but remember: we have had the benefit of a number of scientific revolutions since the time of Laplace. For Laplace's contemporaries it was, in fact, a reasonable hypothesis to make based on a fair reading of Newtonian mechanics. Newton had told us that forces on an object determine that object's acceleration. Furthermore, he had introduced universal gravitation and thus shown how to calculate those forces; they depended only on the relative positions of massive objects and the masses involved. So if the positions of all the bodies were known, so would be the forces they exerted on each other (by the law of gravitation), and thus their acceleration could be calculated (by the second law). Once we have all the accelerations, we may remember from calculus (which Newton also invented) that we can integrate twice to find all the instantaneous positions of the bodies for all times. All that's needed to actually do the integration are the initial conditions of starting locations and starting velocities.

We know now that this interpretation of Newton's laws is too simple. Even in a physical world described only by Newton's three laws and gravity, Laplacian determinism remains an elusive dream. But the reasoning is subtle and had to await the arrival of chaos theory in the 1960s. From a modern perspective, Newtonian mechanics in all but the simplest scenarios actually gives rise to chaotic dynamics that cannot be predicted indefinitely far out into the future. We now speak of deterministic chaos and extreme sensitivity to initial conditions. But this more nuanced scientific view took time to develop, relying to a remarkable extent on the arrival of computers and numerical simulations [18].

So it is not surprising that in nineteenth century Europe determinism was the philosophical message that people thought Newton's description of the universe supported. And with this rise of determinism came its cousin, reductionism. The seed for reductionism is, upon close inspection, already contained in the quote above. We hear of "the tiniest atom" in conjunction with "the greatest bodies." It is now only a small step to eliminate the latter altogether and focus only on the former. Why? Because the greatest bodies are in fact made up of a bunch of tiny atoms; they are, after all, just vast accumulations of tiny particles. So why not restate things even more succinctly and refer only to the forces and motions of the tiniest atoms. Welcome to reductionism—we have arrived!

If strict determinism along the lines of Laplace's quote has since lost much of its luster in the public eye, the specter of reductionism has not receded. One can make the case that it is more firmly rooted in our thinking now than ever. With the rise of the biological sciences, reductionism gave birth to a naive *genetic determinism*. With the rise of neuroscience, it further expanded its reach to include things like mind and consciousness that would not have been subject to it in the nineteenth century. And contrary to the turning of the the tide against strict reductionism within many branches of science, most people are as convinced as ever that science favors this philosophical view, that modern science indeed demands it.

Couple that impression with the overall ascent of science since the time of Laplace, and the outcome is no surprise: riding on the shoulders of science (even as science is now beginning to shake it off), reductionism has gained enormous traction and adherence. Whether consciously or not, it has become an important paradigm through which people make sense of the world around them, and which informs many of their beliefs. It is through its lens that phenomena even outside of the strict purview of science are increasingly viewed. In this way, it has outgrown science itself. The perceived stamp of approval that science lends to reductionism has significantly advanced its public credibility. One is reminded of the words of Émile Durkheim, often considered the father of sociology, when he lamented this state of affairs in 1912.

> Today it is generally sufficient that (truths or ideas) bear the stamp of science to receive a sort of privileged credit, because we have faith in science. But this faith does not differ essentially from religious faith. In the last resort, the value which we attribute to science depends upon the idea which we collectively form of its nature and role in life; that is as much as to say that it expresses a state of public opinion. In all social life, in fact, science rests upon opinion. [19]

One goal of this book is to lay out the scientific case against popular notions of reductionism, and begin to bring perceptions and "opinion" of science back into closer alignment with science itself. But more broadly, the following chapters aim to develop the concept of emergence by illustrating how it manifests within different areas of the physical sciences and beyond. As I will argue, different scientific contexts contribute unique facets to our understanding of emergence. Only through appreciating the nuances of these various reflections and perspectives can we gain a fuller picture of what emergence implies. So let's get started. And what better place to start than in the realm of the "tiniest atom."

Chapter 2
Venturing into the Microcosm

2.1 Quantum Worlds

Everybody seems to love quantum mechanics these days. Whether it's the uncertainty principle, quantum teleportation, tunneling, parallel universes, entanglement, Schrödinger's cat (poor thing)—buzz words from quantum physics abound in the popular imagination, in the works of science fiction, and in new-age inspired magazines and movies. For whatever reason, quantum physics appears to have struck a collective nerve. Within the physics community, in turn, this rare stardom enjoyed by one of its jewels has elicited the full gamut of emotions, ranging from opportunistic embrace to outright disgust. I myself confess to rather mixed emotions upon watching a movie like "What the Bleep do we Know."

Formally, quantum mechanics refers to the fairly dry and abstract physical theory that aims to describes the microcosm—the strange and entirely unfamiliar world at the scale of molecules and below. And it describes processes in this realm consistently in line with actual observations, and extremely accurately so. In fact, its success really began before it had even been fully formulated in its present form.

Classical physics had reached the end of its ropes in making sense of the behavior of atoms by the beginning of the twentieth century, and it began to dawn on physicists that some new approach was needed.

© Springer International Publishing AG 2017
L.Q. English, *There Is No Theory of Everything*,
DOI 10.1007/978-3-319-59150-6_2

Max Planck kicked things off with a new explanation of the so-called blackbody spectrum (the light that is given off by thermal objects) which involved the use of quanta in order to tame the so-called "ultra-violet catastrophe." Neils Bohr followed with a model of the hydrogen atom where the electron orbits were taken to be quantized, thus avoiding the insane prediction by classical physics that the atom should be unstable and destroy itself. Bohr assumed that a quantity called angular momentum of the electron's orbit around the nucleus came in smallest indivisible units. Starting with this assumption, he not only saved hydrogen from self-destructing but was also able to predict the spectral lines of hydrogen to very good accuracy—a remarkable quantitative achievement. Later Erwin Schrödinger found a more complete solution to the same problem of the electron's state inside the Hydrogen atom. Things were finally starting to come together in a major flurry of creative energy in the 1920s. A lot of observations that had remained mysterious were all of a sudden predictable from the new quantum theory.

Fast forward another two decades, and the development of quantum electrodynamics (QED), the first quantum field theory, brought about even more spectacular accuracy of prediction. As related in Richard Feynman's book *QED* [20], one of the first great triumphs of this new theory was that it could match the measured value for the electron's magnetic dipole moment. To give you an idea of the accuracy now achieved, measured values of the electron's "g-factor" and its QED prediction agree to about one part in a billion [21]. What other measurement has that kind of accuracy? To put it into the every-day context, it would be like measuring the length of your finger to within the radius of a hydrogen atom, or the distance from here to the sun to within one and a half football fields. Quantum field theory applied to problems in condensed matter physics, such as the quantum Hall effect, has had similar success [5].

As a result of this spectacular success in matching experimental data, there is little doubt that quantum theory correctly describes the world at the smallest scales. Nothing else even comes close. But how small is small? The answer to that question is remarkably fuzzy. Certainly atoms and simple molecules are small enough to fall under the quantum reign, but what about macro-molecules, molecular motors, viruses, miniaturized transistors on a wafer, or micrometer sized metal flakes? Where does quantum physics end and the classical world start to emerge? The answer is not clear cut, and it depends to some degree also on the experimental context. For extremely precise measurements, where environmental conditions are highly controlled and noise of all kinds suppressed, even macroscopic objects can be made to act according to quantum mechanics. The vast experimental effort that detected gravitational waves for the first time ever in 2016, and going by the name of LIGO, required a truly unheard-of degree of control over their mirrors making up the laser interferometer. This was achieved via cooling the suspended mirrors down to a temperature near absolute zero. Remarkably, what happened was that these mirrors, coming in at roughly 10 kg (or a little over 20 pounds), displayed quantum behavior [22]. Such effects are the rare exception, however, and they usually require experimental efforts bordering on the heroic. Usually, the quantum world is confined to length scales substantially below a micrometer (the size of a typical bacteria).

Now before we delve any further into the strange world of atoms, why is quantum physics even relevant to a discussion on reduction and emergence? What specifically does it contribute? The quick answer is that reductionists always point to the atomic or subatomic realm as holding the final answers for why things are the way they are on any scale. So it seems logical to investigate how things actually exist in that realm according to our best theory—quantum mechanics. And what we discover does not comport with our intuitive picture of elementary particles. In fact, our investigation will suggest that nothing in the microcosm can function as the solid foundation reductionists hoped to find there for the purpose of building everything else from the ground up. Instead, what we discover in many ways will point exactly in the opposite direction.

Truth be told, in terms of interpreting how things exist, quantum mechanics is often frustrating for students and instructors alike. It does not provide us with conceptual pictures or with models we can visualize. In the rest of physics, we have a conceptual layer that allows us to mentally transition from the phenomena we have observed to the mathematical description or representation of the same.

So, for example, let's say you are a student in an introductory physics class and you have just hooked up a coil made of 100 turns of copper wire to a power supply that provides current. You turn the knob and feed current through the coil. Nearby you have a little compass set up. As you crank up the current, you notice that the compass needle is deflected. You bring the current back to zero, and sure enough the compass needle turns back to North; ramping up the current again, it turns away from North just as before. That's the phenomenon you wish to explain. If you immediately had to delve into mathematical equations and formalism, and if you're anything like most people, you would be dismayed or even disheartened. (This is possibly why so many people dislike physics in high-school, when the subject is presented as one long string of equations to be memorized and algebraically manipulated.) No, most of us would like to have a working picture of why this happens.

Fortunately, in this case there is one to be had. For instance, we can say that the current in the coil produces a magnetic field that spills out into the space around it, and this magnetic field then interacts with the compass needle, which is itself a little magnet, creating a torque on it. This is something we can visualize. In fact, the fields can be represented by lines permeating space in a predictable manner (Fig. 2.1).

In quantum mechanics we do not really have this conceptual or pictorial layer at our disposal, and we are asked to leap from the mathematical representation all the way to the level of the experimental phenomenology with nothing in between [23]. In terms of interpretation then, we get nothing positive to cling to. We do get plenty of negative, though. We are "told" how things do not exist. It is this negation that makes quantum mechanics so essential in our discussion of emergence.

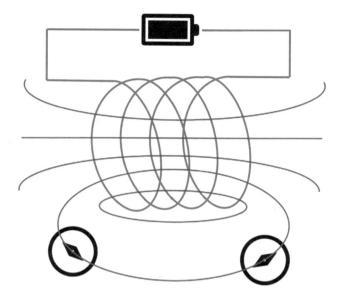

Fig. 2.1 The pictorial model of why the compass needle deflects when a current is set up in a nearby coil. The magnetic field permeating space serves as the conceptual layer to explain the observed phenomenon.

But first, what specifically does quantum mechanics negate? Simply put, it negates the notion of elementary particles existing as solid, point-like entities that have immutable and definitive properties. If I asked you to think of the molecules in the air that you breathe or the atoms in the chair you are sitting in, my guess is you would imagine them as built up from solid little spheres, the elementary particles, kind of like the ball-and-stick models we have seen in chemistry class. You perhaps envision these solid spheres as completely unbreakable, as unaffected in any way to outside tampering. But in the back of your mind, a nagging question now arises: if these spheres are completely solid and immutable, how then can they interact with one another and form atoms and molecules? Doesn't an interaction mean that the partners do something to each other? So maybe we have to revise our generic picture of elementary particles as immutable and solid spheres.

And indeed, in the standard quantum theory, many (but not all) measurable properties of an elementary particle are not to be thought of as immutable or intrinsic, but rather as dependent on the interactions of the particle and on the observational context. (Incidentally, in the contemporary successor theory—quantum field theory—even those properties that were treated as intrinsic in standard quantum theory, such as charge and mass, are now seen as arising from, or at least renormalized by, interactions with other particles/fields.) So, for instance, an electron can sometimes appear to behave like a point-particle, sometimes as a wave oscillating over very large spatial scales, and sometimes as a cloud surrounding a nucleus. Due to this flexibility in appearance of an elementary particle, the term "particle" itself is perhaps misleading and some authors of textbooks thus prefer to simply call them *quantons* [23].

Point, wave, cloud—these are, of course, just mental crutches, and we are warned in quantum mechanics against taking them too literally. The reason is that the *wavefunction*—the mathematical description of the state of a quanton—has no direct physical interpretation. It is not clear what, if anything, the wavefunction corresponds to in the physical world. Some have even argued that the wavefunction encodes only our stochastic knowledge of the state of the quanton, thoroughly lacking any objective reality [24]. So when we say that the electron is in a cloud-like state, we have no real answer for the logical next question: "A cloud of what?" It is questions like this one that can make teaching a first course in quantum theory a little anxiety-inducing.

2.2 Which Way Did It Go?

To appreciate the negation a bit more concretely, let me describe for you an actual experiment that we perform with our students here at the college in their sophomore year of the physics curriculum—the single-photon interference experiment [25]. Using a special setup (the details of which are not important here), we first demonstrate that we can produce single photons—indivisible particles of light. The way we accomplish this feat is by starting with a fairly powerful blue diode

laser. A photon from this laser strikes a special crystal, and what comes
out is two identical photons in the infrared spectrum. We have created
photon twins. Now we feed one the twin photons into a beam-splitter—
an optical device that ordinarily both reflects (say, at 90 degrees) and
transmits some of the light incident upon it. So if you shine a laser beam
on a beam splitter, you end up with two beams—one continuing in the
same direction and the other coming off at a right angle, each getting
half the original intensity.

Now imagine a single photon hitting this beam-splitter, however. If it
really is an indivisible particle of light, then upon encountering a beam-
splitter it must now choose whether to go through it, or be reflected by
it. It can't do both. We can test this by having sensitive light detectors
waiting in both the transmitted and reflected directions and temporally
comparing the detection events that are recorded by both. We find that,
voila, we get no coincidences in time; the two detectors never go off at
the same time. (In reality, there are some random coincidences, but they
are rare in a way that can be rigorously quantified.) This means that the
photons do not divide, with parts traveling to both detectors, arriving
there at the same time.

Having established the photon as an indivisible particle in this specific experimental context, things get interesting. One modification we
could attempt with this particle is to direct it (using mirrors) into what
is called an interferometer. This modified experiment is schematically
depicted in Fig. 2.2; the beamsplitters are labeled BS. An interferometer
basically is a device that presents the photon with two separate paths
to get from A to B, from the input to the output of this device. But the
lengths of the two paths don't have to be exactly the same, and in fact the
path-length difference can be controllably varied in an interferometer.

Let's back up just a bit. These interferometers are not novel con-
traptions; they have been around for a while, and in fact they played
an important part in a number of consequential scientific discoveries.
Perhaps most famously, interferometry was at the heart of Michelson
and Morley's experimental proof of the absence of an ether—an
observation only resolved by Einstein's theory of relativity. The basic

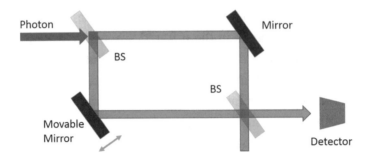

Fig. 2.2 A single photon enters a Mach-Zender interferometer that presents two possible paths to reach the detector. The path-length of one of these paths can be finely controlled via a movable mirror. BS stands for beam-splitter, which can either pass a photon through or reflect it. When the movable mirror is moved, a difference in the lengths of the two paths (leading from source to the detector) is introduced. For waves this usually causes interference, but what happens for a single photon?

idea is that when a wave enters an interferometer, what happens is that the wave splits up (as waves are apt to do), and one component travels one way and the other component travels the second way. When the two wave components recombine at the output channel, several things can happen. If the paths were of identical lengths, then the wave components recombine in phase to produce a large oscillation. If however, the path-lengths are sufficiently different, i.e. if they differ by half a wavelength of the wave, then the components recombine out-of-phase, and they will cancel each other out. The upshot is that as the path-length difference is continuously increased from zero, we observe constructive interference (bright spot), later destructive interference (dark spot), followed again by constructive interference. This interference behavior is well known and can be observed in many different experimental settings for all waves, whether it be light, sound, or waves on the surface of water.

But what if we let a single photon enter such an interferometer. Our first expectation might be that the photon, being an indivisible particle, must chose one or the other path. In that case, we would expect to find that the path-length difference has no effect whatsoever; after all, both paths lead to the output channel where a detector is waiting for the photon. So no matter which path the photon happens to choose, it will get to the output and be registered there by the detector. Tweaking the length of one of the paths does not alter this scenario, and should therefore not matter much, because the photon still gets to the detector. Things are pretty clear cut.

Or are they? It turns out that this expectation is contradicted by observation when we carry out just this experiment. Again, the path-length difference is stepped up, and for each situation, the detector records a certain number of photons per time interval. When these numbers are plotted, we get a sinusoidal graph of bright and dark bands—the interference pattern is back! When we double-check the experiment for errors, we find none; the photons are still entering the interferometer one at a time. Yet, we get this interference pattern that is sensitive to the difference in path-lengths. In order to be sensitive to this difference between the two paths, in a sense the single photon must know about the existence of both of the paths, not just one. Does this mean that the photon must have sampled both paths? Indeed, the strong implication is that of the *self-interference* of a single photon.

So, we are puzzled and we perhaps ask, "What led us to the idea of an indivisible photon in the first place?" Now it doesn't seem so indivisible anymore. But wait, there is one slight modification we can make to this latest experiment that will destroy the interference pattern and restore the conceptual validity of the indivisible photon. All we need to do is try to find out on which path the single photon actually traveled to get from input to output. In this experiment, finding out about the path actually traveled can be done by placing into one arm of the interferometer something called a *half-wave plate*. A half-wave plate is really just a piece of mica, say, and what it does is to change the polarization of the photon going through it. Don't worry about the precise meaning of polarization at this point—just think of it as another property of a photon, just like color is also a photon property. Other than the change in polarization, the photon is not affected by the half-wave plate. But notice that we only put this piece of mica into one of the two possible paths. So by this clever maneuver, now if the photon travels on one path its polarization changes, but if it goes on the other path, the polarization is unchanged. The idea is that we could later at the output see which polarization the photon has and therefore infer which way it went through the interferometer. Aren't we clever! We say formally that we encoded the path information onto the polarization state of the photon.

So let's perform this modified experiment. What do we see? No interference pattern as the path-length difference is changed! The detector counts are not affected by the mirror motion, and they are therefore independent of the difference in path-lengths. Simply by adding that one half-wave plate made the photons behave like indivisible particles again that travel one way or the other, but not both. What's interesting here is that we didn't even attempt to ascertain the polarization state of the photons at the output. It seems to be enough that we could do this in principle. This is really a remarkable result, if you think about it.

What we have discovered here for single photons is actually a general feature of quantum mechanics and holds not just for photons but for all quantum particles (or quantons). When we give a particle a choice between different ways of getting from A to B, and we don't look which way is actually realized, it chooses all of them. This is essentially the Feynman path-integral formulation of quantum mechanics; we have to sum over all possible paths. However, if we indirectly try to find out something about the paths actually taken, then the particle will take one specific path (at random, but with certain well-known probabilities). For the double-slit experiment with electrons, Feynman put it thus [26]:

> It is true, or isn't it true, that the electron either goes through hole 1 or it goes through hole 2? [...] What we must say (to avoid making wrong predictions) is the following. If one looks at the holes, or more accurately, if one has a piece of apparatus which is capable of determining whether electrons go through hole 1 or hole 2, then one can say that it goes either through hole 1 or through hole 2. But, when one does not try to tell which way the electron goes [...] then one may not say that the electron goes either through hole 1 or hole 2. [...] This is the logical tightrope we must walk if we wish to describe nature successfully.

Notice Feynman's careful phrasing at the end: "[...] then one may not say that [...]." There it is again—quantum mechanics negating one of our mental constructs, telling us how *not* to think about the situation.

We can make one more slight modification to our experimental setup from before—one which really underscores the subtleties involved, and one that typically perplexes my students the most. Having encoded the path information in the polarization state of the photon, let's see what would happen if before the detector at the output, we scrambled that polarization state so that it could not yield path information. This is suggestively called the *quantum eraser* experiment. The way we can practically do this is simply by placing a polarizer at 45°. Mind you, this polarizer is placed just before the final detector, so that the photon has already gone through the interferometer and arrived at the output before it encounters the "scrambling" polarizer. Could this polarizer now make any difference? This is the truly baffling aspect of quantum behavior—it does make a difference in that the interference pattern comes back!

It is difficult to make sense of this result. The photon has already left the interferometer before it encounters this modification; one might have guessed that it would have already "decided" if it wants to "split up" and sample both paths or "stay together" and sample only one path. But that is not how nature works. All we can conclude is that the overriding principle that determines whether we see single-photon interference or not is whether it is possible for us to know the path traveled; if no, then we get interference, if yes, then we don't. It doesn't matter how, when or where we obtained the path information. We just cannot trick the photon like that.

One of the larger take-away lessons from this experiment (and many others) is that particles don't necessarily have to be here or there; they can reside in so-called *superposition states* of these base states. For instance, if a quanton (electron, proton, etc.) has a choice between two spatial locations, then in general the quanton will be in a superposition of those positions.

Another concrete example is the ammonia molecule, NH_3, in which the nitrogen atom can be on either side of the plane spanned by the three

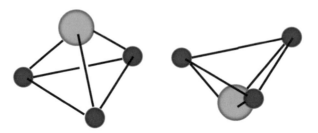

Fig. 2.3 The ammonia molecule is really in a superposition state of the two configurations shown. The Nitrogen atom is both on *top* and on the *bottom* of the hydrogen plane

hydrogen atoms, as shown in Fig. 2.3. In reality, however, the nitrogen is on both sides at the same time—something that is, of course, hard to draw or visualize—hence the pyramidal structure that we usually identify with ammonia. In general, only upon measuring the atom's whereabouts does it get pinned down to a definite location. Since the result of a measurement has to return a definite position, the atom complies and "chooses" one place or the other. We say, its wavefunction (the superposition state) *collapses* to one of the base states.

It would be tempting to interpret the superposition state as reflecting only our ignorance about the whereabouts of the particle before a measurement is taken. In this view, the measurement did nothing to the particle at all, it simply clarified for us where it was sitting all along. This interpretation, however, is untenable and has been rejected, as it contradicts experiments designed to test this difference in interpretation (Bell's inequality) [27–29]. Even in our single-photon interference experiment, this view does not give us the correct answer. How could the photon interfere with itself if it were really either here or there?

Instead, we have to say that elementary particles, quantons, generally do not have definitive properties such as location in space, but also spin orientation, momentum, and such, before a measurement takes place. When a measurement is taken, the wavefunction collapses into a base state (called *eigenstate*) of the observable being measured. In this state, the quanton then has a definitive value of the property being measured.

But what's interesting is that in that state, it loses all definity in other observables. This is the famous Heisenberg *uncertainty principle*. We cannot know both the position of a particle and its momentum. If we have pinned the particle down in space, we lose all knowledge of its momentum, and vice versa.

Perhaps you are feeling a bit woozy. Don't worry, quantum mechanics has that effect on people. And after the first dizziness has lifted, there are typically a couple of reactions people have. Excitement at the possibility that things might actually be different than how we always imagined them, is one reaction. Shock followed by staunch denial is another. I remember when I first learned about the outlines of the quantum picture in high school, I didn't want it to be true. I wanted it to go away, to be disproved. It seemed so different from everything I liked about physics—the predictability, the orderliness, the deterministic regularity. As it turns out, I was in excellent company. Einstein never warmed up to quantum physics, maintaining until his death that it would eventually be supplanted by a better theory. Even Schrödinger later distanced himself from the implications of the theory he himself had been instrumental in formulating. In college, when I first read the often contentious correspondence between Einstein and Bohr over the meaning of quantum physics, I still always instinctively rooted for Einstein (although I have changed my mind since).

Someone who is not that much into science may have a third, less emotional, reaction. Some people I talk to just don't think that anything in science has to be taken that seriously. The attitude is summed up in the statement: "Theories come and go. Who's to say this is the final word." Science always has a certain shelf life, the thinking goes. It is true, of course, that it is much easier to disprove something in science than it is to prove it. In fact, an absolute proof (in the sense mathematicians use the word) is not possible at all. However, if over the course of decades people as smart as Einstein have tried to poke holes in and tear down quantum physics, and if generations of experimentalists have tried the same, and the theory is still standing intact nonetheless, then our confidence in it dramatically increases. The fact is that quantum physics just seems to work every time, no matter how precise the test to which we subject it. It is simply not easy to dislodge. Physicists are now quite sure that the uncertainty principle, for instance, will always be a part of any future successor theory.

2.3   Entanglement and Irreducibility

It thus seems that we have to take quantum physics seriously, however grudgingly. Unfortunately, things get stranger still. We have not yet exhausted the strange predictions of quantum mechanics by a long shot, and a new level of weirdness comes when we consider more than one quantum particle. Just earlier we learned that each particle by itself can be in a superposition state, but what about two of them together? If the first particle is in a superposition state of A and B, and the second particle in a superposition state of C and D, would we not expect the two-particle system to be in a superposition state of AC, AD, BC, and BD? In other words there should be four possibilities of the combined system—these are the four possible combinations that the two-particle system could be found in.

It's roughly akin to flipping two coins. You launch the first coin into the air, it lands on the table, but before you look, you have your hand covering the coin. It could be heads (H) or tails (T), but you don't yet know which. Now you flip the second coin in a similar manner using your other hand. Your two hands now conceal the state of the two coins. We have four possibilities: HH, HT, TH, or TT, one of which will be revealed as soon as you remove your hands. The difference in the quantum example is that the system does not have to choose one of those four possibilities, but could be in a superposition state of all four. And indeed, such a superposition state of all four base states generally would be allowed and can be set up (with certain restrictions for two particles of the same kind).

However, in addition to the aforementioned, there are other combined (two-particle) states that are harder to interpret. For instance, what about the superposition state of AD and BC alone? Both particles can still be found in either of their two states (A and B for the first particle, C and D for the second). But it is conditioned on the state of the other particle and not independent of it. For instance, if and only if particle 1 is in state A, is particle 2 in state D. We call such a situation an *entangled state*. Returning for a moment to our coin analogy, it would be as if instead of the four possible states, the coins could only choose two of them, say, HT or TH. By some strange magic, whatever state the first coin is in, the other coin has to be in the opposite state. If the first one is heads, then the second one has to be tails, and vice versa. This is the truly puzzling aspect of entanglement: We cannot think of such a state as particle 1 residing in this superposition state and particle 2 residing in that superposition state. The two particles from a genuinely new and *irreducible* two-particle state.

It's worth exploring these entangled states in a bit more detail, if just to get some better 'feel' for them, but also because they hold deep relevance to our discussion of emergence, as we will see. If you do not like equations, feel free to skip ahead to the next section where you can seamlessly rejoin.

Ready? For concreteness, let's say that the spins of two electrons have become entangled, which means that they both partake in the two-electron entangled state. For each electron, its spin can be either up or down, or it can be in a superposition of up and down. We will represent these states with little arrows pointing either up or down. It is also customary to surround those arrows with brackets. So the spin of electron 1 pointing down would be represented as $|\downarrow\rangle_1$. But now we examine the following state (which, technically, would still need to be *normalized*):

$$|\uparrow\rangle_1 |\downarrow\rangle_2 + |\downarrow\rangle_1 |\uparrow\rangle_2 \qquad (2.1)$$

Fig. 2.4 An example of an entangled state. The spins of two electrons have been entangled by an event that happened at the origin. Now they are flying rapidly apart, but the system is in a superposition state of the two pictures

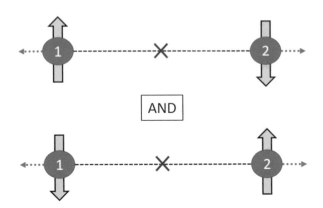

The subscript around the brackets denote the particle to which we are referring, and the sum means that we are in a superposition of two basic states. In Fig. 2.4, I have made the attempt of representing this state visually. The funny thing is that since we have entangled only one property of the electron, namely spin, this does not dictate most aspects of their spatial information. In fact, we can set things up so that once these electrons have been entangled (via some process yet to be specified), they can fly apart and after some time be separated by long distances from one another. We have arrived at the famous *EPR paradox*, named after Einstein, Podolsky and Rosen, who first formulated it with the intention of showcasing the absurdity of quantum mechanics [30]. In the end, what they really accomplished was to deepen our appreciation for the counter-intuitive subtlety of quantum mechanics and the world at the smallest scales. But what is the paradox?

As we have said before, when a measurement is performed on a quantum system, its superposition state collapses. So let's say that we were to perform a spin measurement on one of the electrons, the first one, and that the result were 'up'. Which part of the superposition state is consistent with the first electron having spin up, or $|\uparrow\rangle_1$? The first part! And so the two-particle state must collapse to:

$$|\uparrow\rangle_1 |\downarrow\rangle_2 \tag{2.2}$$

In essence, the measurement just eliminated the lower half of Fig. 2.4. What consequence does this collapse have on the second electron on which no measurement was performed? Examination of the collapsed two-electron state reveals that it too collapsed, namely to the spin-down state, or $|\downarrow\rangle_2$. This is what Einstein called "spooky action at a distance." The second electron's state instantly changed in response to a measurement on the first electron even though the two are perhaps light-years away from one another. However 'spooky' this situation may seem, the phenomenon is by now a well-established experimental fact [29, 31]. Particles may be far separated in space, and yet remain entangled in one of their properties. In a sense, the particles' separate identities are subordinate to the common two-particle state in which both partake. The separate identities of these two electrons is secondary to their relationship with one another as expressed by the entangled state.

2.4   Quantum Emergence

We now return to the broader question of how these lessons from quantum mechanics inform our understanding of emergence. Before focusing on entanglement and the quantum physics of many particles, let's first look at the quantum notion of elementary particle again. Is it relevant to the debate between reduction and emergence? Yes, indeed! In the reductionist paradigm, phenomena at higher levels of organization or complexity are just apparent phenomena, they are not substantial or real, and they certainly do not possess causal power, however conceived. They are mere 'epiphenomena'. The true causal power, according to this view, is vested at the smallest scales in the elementary particles, since presumably they have intrinsic properties that lend them this inherent causal power.

The only problem is that this view of elementary particles is wrong and has been thoroughly rejected by physics, as we have just seen. Quantum mechanics says that this is exactly not how things exist at the smallest scale. Instead of definity and solidity, we find only indeterminacy and potentiality, embodied in the wavefunction and superposition states; we find only contextual properties that depend on experimental conditions. Sometimes the particle really travels on a particular path to get from A to B, and sometimes it chooses multiple paths.

If we look at quantum field theory, the successor theory to traditional quantum mechanics, things are very similar. Here particles are interpreted as discrete excitations of the corresponding field, so photons are quantized excitations of the electromagnetic field, for instance. But now particle properties are understood to arise due to interactions with other fields. It is the interaction with the electromagnetic field, in the form of absorption and emission of photons, that "bestows" charge on an elementary particle. Or put another way, it's a defining characteristics of charged particles to "couple" to the electromagnetic field. It is the interaction with the Higgs field that bestows mass. The properties of elementary particles are seen as deeply relational.

Therefore, from the quantum perspective, the sharp distinction invoked by reductionists between the bottom and the top dissolves. It is not true that only at the higher levels do things become configurational and relational, dependent on other factors outside of themselves, dependent on their parts and their relationships. That things are relational and insubstantial is already true at the lowest level as well, even if not for exactly the same reasons or in the same way. As philosophers Campell and Bickhard put it [8]:

> There is no 'bottoming out' level in quantum field theory - it is patterns of process all the way down, *and* all the way up. [. . .] If being configurational makes a property or power epiphenomenal, then everything is an epiphenomenon.

Before we discuss entanglement and what it tells us about emergence, let me illustrate the situation so far in a more colorful way. So imagine it is your birthday and your kids, knowing your fascination with chemistry, have given you a ball-and-stick kit as a present, so that you can build your own molecules. You are excited and start building some basic molecules; perhaps you feel inspired to build benzene, the carbon hexagon with the hydrogen atoms sticking out. Looking at the molecule you just built, you admire its symmetry, its compact structure. Even so, it remains obvious to you that the balls and sticks are the basic building blocks, and that the molecule was simply constructed from them. At the end of the day, when you have disassembled everything and put things back in the box, what remains is the balls and sticks. It so happens that the box has a place where the sticks go and a place where the balls go, with a divider between them. The molecule only existed as the relationship between these basic building blocks.

So far so good, but now your box suddenly gets zapped and quantum mechanics has come alive in your molecule kit. You open the box again and look inside, and you don't recognize any of the parts from before. There is nothing in the box that resembles a stick, nothing that looks like a ball. Instead, things seem oddly smeared out over the two sides of the box. On the left side, where you had put the sticks, you see wave-like patterns of different wavelengths; some of the items with the shortest wavelengths appear even to be leaking out of the box a little and spilling over into the right side of the box, where there are also weird wave-like patterns. You reach inside, but there is nothing solid to grab onto; you can't get the stuff out of the box that way.

All of a sudden, you see a flash of light emanating from the the box. You look and notice something funny—on the left side of the "sticks" one wave-pattern that before was protruding across the divider a bit is now completely missing; it seems that it has hopped over to the right side of the box, where it has caused a significant alteration of the wavelike patterns. Presently, in order to get the stuff outside the box, you have the bright idea of simply flattening out the sides of the box. When you do that, some of the entities seem to diffuse out in all directions like mist, but all of a sudden something new pops into existence. Before you there is now one 'object' with some recognizable structure that somehow spontaneously assembled itself. Gazing at the object, you find that it vaguely resembles two balls and a stick between them—but only vaguely. When you try to reach for the stick, your hand goes right through it; the same is true for the tiny balls. They are like clouds of recognizable shape, but with no sharp edges. Bemused and puzzled you scratch you head and think: "What is this molecule made up of? Is the whole really defined by the parts, or is it the other way around? Does the whole, in fact, define the parts? The isolated parts did not resemble anything like sticks and balls, after all. Only now in this molecular configuration is there this vague resemblance." You wake up and quickly realize that you had dozed off on the floor, your kit sitting right beside you. It was just a dream. But on the atomic scale this is not a dream or fantasy, this is our best description of reality!

Let's return for a moment to the case of entangled particles. Remember, the hallmark of an entangled state is that it is irreducible to a product of single-particle states. In more familiar language, two particles that have become entangled via a prior interaction form a two-particle system that cannot be thought of as particle 1 being in some state AND particle 2 being in some other state, separately. That doesn't work. So the entangled pair is already an emergent phenomenon, in that it is irreducible to single-particle states. If we treat the isolated, single particles (as well as their states) as level 1, then there is no description within level 1 that accounts for the state of the entangled pair. That description occurs at level 2, at the two-particle level. Erwin Schrödinger recognized the philosophical implications that entanglement presented early on, when he wrote in 1935 [33]:

> Best possible knowledge of a whole does not necessarily include the same for its parts. [...] The whole is in a definite state, the parts taken individually are not.

It is of course true that when we have to write down the definite state of the whole, we have to, notationally at least, refer to the parts. So for instance in Eq. (2.1), the subscripts 1 and 2 to the right of each bracket enumerate the specific particle that becomes part of this state. We cannot entirely get away from these labels. Nevertheless, one could argue that these are vestiges of the former isolated particles that in this entangled state have ceased to exist the moment they became entangled. This comes close to the idea of "fusion" proposed by philosopher Paul Humphreys [10]. Here the constituent parts in the process of becoming part of the whole give up their individuality altogether and fuse into the collective state.

In this view, the properties of the whole and processes at this level cannot be reduced to properties or processes of the parts by default, since the parts themselves have disappeared once they aggregated to form the whole. Others in philosophy have questioned to what extent we can say that the parts have completely disappeared in an ontological sense [34]; however, the idea of fusion does evoke quantum phase transitions that we will examine in the next chapter. And since these phase transitions do produce states of matter, like superfluids and superconductors, that have characteristics so unlike ordinary matter, the practical importance of entanglement cannot be denied. The relationship between parts and whole is re-negotiated, and this has very measurable physical consequences.

A couple of additional observations about entanglement bear mentioning. The first is that, of course, more than two particles can become mutually entangled. We would obtain states like in Eq. (2.1), but with another bracket of subscript 3 in the product of each term (and more such terms). These particles are generally allowed to access more than two states, and therefore we can form superposition states involving more than just two base states (such as spin up and spin down). So things are fairly straightforward to generalize along these lines. Secondly, not just the states of particles, but also their time evolutions can sometimes become entangled [34].

Usually, we think of a system comprised of parts to be defined by the states of all the individual parts; in this sense, we speak of the whole as the sum of its parts. If I have a collection of papers and books on my desk, as I have them right now, I think of the state of my desk as given by the states of every paper and book put together: the location of every book and paper on the desk, their individual orientations, colors, etc. The desk is comprised of all these things on it, and so the desk state should also be the sum total of the individual states. What we have with entanglement in quantum physics is an example where the roles are reversed, however. The state of the whole is explicitly not given by an enumeration of the individual states. The state of the whole has primacy over the states of the parts.

One caveat is, of course, that we are dealing with quantum mechanics here. There is no analog of entanglement in the classical world. It would strike me as more than weird if my desk were not reducible to the states of its parts, like the papers and books on it. Even if I allowed for interactions between objects on my desk, like my papers folding themselves into paper animals and coming alive to fight out an epic battle on my desk, even then would I still say that at any given time, the state of things on my desk is a direct function of the position, velocity, orientation and rotational motion of each of the paper animals. Nonetheless, quantum entanglement does exist in nature and has measurable consequences, and so cannot be dismissed even if there is no analog of it in Newtonian mechanics and the other sciences. The macroscopic weirdness of a superfluid—the fact that it can creep up the sides of containers, or the absence of buoyancy—is a consequence of massive degree of quantum entanglement between the microscopic contituents. Couldn't that be construed as a mild form of reductionism, you may ask? Maybe, but remember that what is happening is the very elimination of the particles' separate identities.

There is one more striking feature of entanglement, and that is what physicists refer to as *nonlocality*. We have already mentioned that the two entangled particles can find themselves separated by large distances in space. As theorist and author Brian Greene points out, usually when things are separated in space, they are considered separate entities by this virtue alone [35]. Two asteroids floating in space are identified as separate objects precisely because there is space separating them; if there weren't any space between them, they would have crashed into one another, in which case they would suddenly form one combined asteroid. This is not to say that separate objects can't interact, of course, but that interaction takes the form of communicating through space in some way. And this communication has to obey the universal speed limit of c—the speed of light, as demanded by Einstein's special relativity.

This, in a nutshell, is meant by 'locality'. Things done here can effect something over there, but only after a sufficient time (not less than an absolute minimum) has passed. In ancient Rome, when something significant (like a rebellion) had happened in the far corners of the empire, the emperor in Rome would not hear of it until weeks later. Nowadays, with radio, satellite TV, and the internet, communication is quicker, but still not faster than the speed of light.

From our earlier discussion, an entangled pair of electrons is to be thought of as one irreducible entity, but that entity can be spatially distributed. Here it seems the rule of thumb that things separated in space ought to have separate identities does not hold true. In fact, when one electron is disturbed by a measurement, say, and changes state, then instantaneously the other electrons also changes state in some sense. The very fact that this change happens instantaneously should suggest to us that this is no ordinary communication between separate objects that have interacted. No, this "spooky action at a distance" is an indication that the whole really does supersede the parts from which it was made.

Chapter 3
The Aggregation of Particles

3.1 The Arrow of Time

Ludwig Boltzmann was a towering figure in the physics of the late nineteenth century. Every undergraduate physics and chemistry major now studies the Maxwell-Boltzmann distribution, Boltzmann statistics, the Boltzmann factor and perhaps also the Boltzmann equation. Despite his significant fame, much of which he witnessed during his own lifetime, he is also one of physics' most tragic protagonists.

His life's work was dedicated to re-developing large swaths of physics (chiefly the field of thermodynamics) from the premise that matter was made up of atoms. While thoroughly uncontroversial today, during Boltzmann's time the atomistic view was still hotly debated in physics and philosophy circles. The majority view could probably be summed up as holding that atoms might have some utility as practical constructs, but that they should not be taken too literally as real constituents of matter. Needless to say, this view contradicted Boltzmann's entire program—a fact that would lead to much professional tension and antagonism.

That's not to say that Boltzmann was a fringe scientist working in isolation. Quite the contrary was true. Many important scientists of the day held him in high esteem and collaborated with him (including Lorentz, Kirchhoff, Helmholtz and Rayleigh), and he had many

© Springer International Publishing AG 2017
L.Q. English, *There Is No Theory of Everything*,
DOI 10.1007/978-3-319-59150-6_3

students who became influential scientists in their own right (including Ehrenfest, Meitner, Nernst) [36]. But he also made a number of fierce enemies, perhaps chief among them the philosopher and physicist Ernst Mach.

Mach is now probably best known for his studies of shock waves, like the ones created when the sound barrier is broken by a bullet or a fighter jet. He also contributed to the study of light in media, which he took to be a continuum, but later in life he began shifting his focus increasingly toward the philosophy of science, where he advocated for a strict positivist interpretation. This lead him to become an outspoken and formidable critic of the atomic hypothesis, immediately bringing him into conflict with Ludwig Boltzmann. When he became Boltzmann's colleague at the University of Vienna, the proximity did not engender friendship or comity. In fact, their relationship mirrored the larger battle raging in physics, creating a bitter and dogmatic schism between adherents of "energetics" and the atomists.

Boltzmann was constitutionally uncomfortable with professional conflict or ideological strife; he thrived in an atmosphere of loose collaboration and friendly competition. But he also knew that he was onto something big, something worth defending against energetics, something that hadn't yet been fully understood, let alone embraced, by the larger scientific community. He had written papers that despite their mathematical sophistication were increasingly being noticed, first in Britain and then in continental Europe and America. He was confident about his ideas, yet simultaneously plagued by self-doubt about whether he could make himself understood in lectures and in writing.

His grand vision, sharp intellect, and deep interest in philosophical interpretation all fueled in him an almost missionary zeal to teach and convince others, but this ambition often collided with the reality that the intellectual landscape was polarized; most people were not yet ready to receive the message. In 1906, after prolonged struggles with physical and mental illness exacerbated by professional conflict and disappointments, Boltzmann took his life while vacationing with his family in Italy. On his gravestone in Vienna, just above the large bust

Fig. 3.1 The grave of Ludwig Boltzmann, 1844–1906. His famous entropy formula is engraved at the *top*. (Image by Martin Roell, https://commons. wikimedia.org/wiki/File: Boltzmann_ (427167382).jpg, licensed under CC BY-SA 2.0)

of his likeness, is carved his famous formula for entropy, $S = k \log W$. Had he lived even just a few years longer, he would likely have come across Einstein's 1905 paper on Brownian motion which contained in it the ultimate vindication of the atomic hypothesis (Fig. 3.1).

What was Boltzmann's great insight, his big idea to which he dedicated his life? In short, it was the first physical explanation for why time flows in one direction only, from past to future, but never in reverse. Boltzmann had discovered nothing less than the arrow of time.

It is a fact that for individual atoms there is no arrow of time. Traditional quantum mechanics is very clear on this point [37, Chap. 6], [38, Chap. 4]. If a scattering process works in one direction, it must also be allowed in the other. A photon may be absorbed by an atom, causing an electronic transition into an excited state, but the reverse phenomenon must be equally likely—a downward electronic transition and an emitted photon. We can say that photon *absorption* and *spontaneous emission* are two sides of the same coin. The famous Schrödinger equation, which governs the evolution of a particle's wavefunction, exhibits time-reversal symmetry (upon simultaneous complex-conjugation).

One could perhaps make the objection that in modern particle physics it is now well established that one of the four fundamental forces—the weak force—does not perfectly respect this symmetry. This curious fact is actually a consequence of another symmetry violation perpetrated by the weak force on rare occasion, the so-called charge-parity (CP) violation, which came as a real shock to the physics community when it was first discovered (see, for instance, [39] for some interesting history). Nonetheless, this violation is strictly limited to weak-force processes, such as certain radioactive decay events. The weak force has an incredibly small spatial range, much smaller in fact than even the nucleus, and it is not of any consequence in the motion of stable atoms.

Even before the arrival of quantum mechanics, people like Boltzmann realized that time as we perceive it did not exist in the microcosm. If you could somehow manage to shoot footage of a few atoms doing their thing, it wouldn't make the slightest difference if you played the video forward or backward. All we would see is Brownian motion.

Not so in our world. Most people can detect that a movie is played backward in a matter of seconds. The reason we can easily tell is not just because people are walking backwards and cars are driving in reverse. These things could actually happen in real life also—actors could be instructed to walk backwards, and so on. There are some processes, however, that couldn't be contrived to run in the opposite direction, some things a movie director couldn't fake other than by literally reversing the order of images after the fact.

Think for instance of boiling water in a pot. You see steam coming off and dispersing throughout the kitchen and beyond. What would the time-reversed footage look like? Water molecules from all across the room gathering back at the pot, rejoining the liquid water in the pot. We would intuitively know something was wrong and almost immediately spot the mistake. Or think about a coffee mug dropping to the floor, spilling its hot liquid and shattering upon impact. Energy is conserved at all times, it is never destroyed but only transformed. So why couldn't the reverse process happen, whereby the ceramic shards reassembled spontaneously, with molecular bonds reforming, the liquid leaping up from the floor and coming together inside the mug? It wouldn't violate the conservation of energy! The answer that was invariably given prior to Boltzmann was "because the second law of thermodynamics prohibits it."

But what was the physical basis for, the mechanism behind, the second law? Where did it come from? It wasn't clear to anyone. And looming even larger was the dichotomy of complete reversibility in the microcosm and irreversibility in the macrocosm. How did this asymmetry arise? It seemed mysterious and was, in fact, referred to as the "Loschmidt paradox" [36], [37, Chap. 3]. In the words of William Thompson (also known as Lord Kelvin),

> If, then, the motion of every particle of matter in the universe were precisely reversed at any instant, the course of nature would be simply reversed for ever after. [...] Heat which had been generated by the friction of solids and dissipated by conduction [...] would come again to the place of contact and throwing the moving body back against the force it had previously yielded. Boulders would recover from the mud the materials required to rebuild them in their previous jagged forms, and would become reunited to their mountain peaks from which they had formerly broken away. [36, p. 98]

Yet, we never observe boulders regaining the kinetic energy they lost during impact as heat, and thus flying back up the mountain to rejoin the peaks from which they had fallen. Why not? It is indeed puzzling that microscopic reversibility should not also imply macroscopic reversibility. After all, matter is made up of nothing but small particles, atoms. There is no secret ingredient, other than atoms, making up large objects (such as boulders) that would bestow on them this different quality. How could large objects behave so differently when they were really nothing other than huge assemblies of small objects?

This circumstance must, of course, remain forever mysterious to strict reductionists. Conversely, we see in this problem another glimpse of emergence. Could it be that irreversibility is not a mechanical law of nature along the lines of Newtonian gravity, but rather an emergent phenomenon that reveals itself only as more and more atoms aggregate and interact with one another? As we shall see, this was where Ludwig

Boltzmann was heading in the late 1860s. His efforts culminated in 1872, when he published a long paper with the title of "Further studies on the thermal equilibrium of gas molecules" [40]—a paper that would become enormously influential despite its formal and somewhat nondescript title.

Boltzmann invited us to ponder a thought experiment roughly along the following lines. Imagine that in one box you have some salt, comprised of many individual grains of salt crystals. In another box, you have some pepper, again made up of many little pepper corns. What would happen if you poured the salt into the pepper box? Easy—the salt would sit on top of the pepper. The white salt layer would be above the black pepper layer with a clear boundary in between. You can probably anticipate what comes next. Yes, let's shake the box. What happens? The answer is quite intuitive—we would expect to find a uniformly gray content. The salt and the pepper have thoroughly mixed. This mixing through stirring is, of course, a basic principle of cooking.

Now let's ask what the reverse process would look like? By continuing to shake the box, could we ever get back to the two segregated layers? Again, intuition tells us, no, that would never happen. No matter how long we shake or stir, the salt and the pepper will never separate themselves out in space again. In fact, common sense tells us that the longer we shake, if anything, the more thoroughly we mix.

So here is the conundrum. If we follow any individual salt crystal in time as the container is shaken, the motion is quite random. It collides with other salt and pepper particles, bounces off of them in all kinds of directions, is flung up and down by the shaking, sometimes hits the container wall, and so forth. If we blocked out everything else, and followed the trajectory of just that specific salt crystal, it could equally well happen in reverse. The particle moving on that same trajectory in reverse order would not look unreasonable. In fact, that trajectory would be equally likely to manifest.

By zooming in on one particle, we lose sight of the larger trend, namely the trend towards greater disorder. Initially, all the salt grains are on top and all the pepper is on the bottom. It is an orderly affair. But as we start shaking the container, that order gradually begins to disappear and gives way to a more disordered and mixed state. Of course, we have to zoom out to even notice the increase of disorder because it is a system property. At the level of the single grain, even the notion of order and disorder loses its meaning.

The second law of thermodynamics says that the disorder, more precisely the entropy, of an isolated system never decreases with time. It is easy to produce the mixed state, but nearly impossible to recover the ordered state once it is gone. Not only did Boltzmann realize what principle lay beneath this law, but he was ready to quantify his insights in mathematical terms. He realized that the second law was essentially about statistical odds. In 1872, he was ready to present his famous H-theorem.

To many of Boltzmann's contemporaries, the H-theorem must have seemed like a sophisticated sleight of hand. Indeed, it can appear that way to students even to this day. Some of us can perhaps follow the steps in the argument that incrementally lead from A to B, but when arriving at point B, we are left feeling that some trick must have been played on us. How else can we explain that out of the raw materials of microscopic reversibility were crafted an equation which is not time-reversible. Upon further thought, it dawns on us that the trick Boltzmann employed came in the form of a statistical analysis of particle interactions [37, Chap. 2].

Simply put, the disordered states are many, and the ordered states few. If you had to select at random the location of every grain of salt and pepper, the likely configuration would be a nicely mixed state. Very rarely would a configuration arise for which all the salt grains were on top of all the pepper corns. Boltzmann was even able to calculate how unlikely that would be—the answer depends on the numbers involved. The larger the numbers of particles, the more unlikely such ordered states become. By the time we get down to gas molecules of, say, oxygen and nitrogen in a small container, the chance of segregation becomes zero for all intents and purposes. When we are dealing with the 10^{20} molecules in my small box, it basically never happens. The reason is not because the system is physically prevented from setting up an ordered state or moving towards greater order, but because the odds of it happening are infinitesimally small. One would have to wait trillions upon trillions of lifetimes of our universe before we would see such a configuration arise spontaneously.

The upshot is that probability dictates that systems, when left to their own devices, never move away from disorder and towards greater order. The entropy never decreases. When an ice-cube is placed on a hot stove-top, heat flows from the stove-top into the ice-cube causing it to melt. Heat does not flow in the other direction. It does not flow from the ice-cube into the stove-top, causing the former to cool further and the latter to become hotter. It is fairly straightforward to prove that heat flowing in the "wrong" direction would increase the order and violate the second law.

Does that mean order can never manifest itself? That doesn't seem right either. And indeed, order can establish itself in accord with the second law, but it has to come at the expense of greater disorder elsewhere. Think of water freezing into ice when the temperature drops. The water molecules become fixed at regular points in space, forming a highly structured lattice. Hasn't the order just increased? Yes, but we have to remember that during the freezing process, latent heat was released into the environment, and that heat created more disorder there.

Thanks to the work of Ludwig Boltzmann and later that of others as well—the great American scientist Josiah Gibbs should be mentioned—we now understand the laws of thermodynamics to be rooted in the statistics of large numbers of interacting particles. The second law of thermodynamics is not a law in the sense of other laws within classical physics. Newton's second law, for instance, states that the acceleration of an object is proportional to the net force acting on it. It cannot be violated by virtue of being a law of nature—no further explanation can be given. Boltzmann showed us that the second law of thermodynamics is different. It is obeyed only because its violation is extremely improbable.

In his quest to make sense of the second law, Boltzmann produced two gems—the Boltzmann equation and the H-theorem. The equation was able to follow any system of particles out of equilibrium along its path from a more ordered configuration to a less ordered one. As mentioned earlier, the remarkable aspect in this context is that the equation is not invariant (or symmetrical) under time reversal. That is an explicit mathematical feature of the equation—letting time run backwards changes the form of the equation. The symmetry is broken. In fact, Boltzmann introduced a quantity, H, that would always decrease with time, never increase.

It is worth reiterating that for this to happen mathematically, Boltzmann could never have started from the equations of motion governing any one molecule. The novel feature of the broken symmetry had to come about by taking a statistical approach to the problem. Loosely speaking, his equation had to zoom out spatially by many orders of magnitude in order to accomplish the feat of irreversibility. It is, therefore, not formally derivable starting from a microscopic description.

Emergence holds that novel phenomena arise within every new layer of scale or complexity. One such emergent phenomenon seems to be the nature of time itself. Perhaps only in a world that is out of thermal equilibrium and where the objects of experience are comprised of an unimaginable number of atoms does it make sense to talk about the arrow of time. The reason we never see particular processes run in reverse can be viewed as one profound consequence of the aggregation of particles. In the words of Ludwig Boltzmann,

> Just as at a certain place on earth we can call "down" the direction towards the center of the earth, so a living being that finds itself in such a world at a certain period of time can define the time direction as going from less probable to more probable states (the former will be the "past", the latter the "future").

Whenever physicists talk about time, what complicates and often confuses matters is the fact that there exist a couple of different notions of time in physics. Einstein's theory of special relativity, for instance, fuses time and space into four-dimensional space-time. Here we learn, despite what our classical intuition tells us, that the time-interval that is found to elapse between one event and another actually depends on our state of motion and is thus not intrinsic to the two events. We also have to give up on the idea of a universal "now"—according to Einstein, two observers will generally disagree about whether two events occurred simultaneously, both being correct from their own side. Despite this very counter-intuitive description of time, nowhere within the theory of special relativity do we get an explanation of the fact that we perceive time to march forward in one direction only. The arrow of time does not naturally arise within Einstein's paradigm.

The general theory of relativity adds gravity into the picture and postulates, for instance, that massive bodies appear to alter the flow of time in their presence. Now the very "fabric" of space-time is predicted to deform in the vicinity of massive objects. Sometime that deformation can be quite drastic, as in the case of black holes. The theory even permits us to envision the possibility of worm-holes—an idea not lost on countless science fiction writers since. (If you've watched the movie *Interstellar*, you may recall that much of the plot rests on the appearance of a worm-hole in our solar system and on the ramification of time dilation near a black hole.) Undoubtedly, general relativity makes some mind-bending predictions but, again, no real hint of a temporal arrow can be discerned.

If we think about it, there is an undeniable biological basis to our perception of time. To illustrate that connection, we can ask ourselves why a nanosecond (or even a microsecond) is too short of a time interval for us to perceive. We cannot follow processes that evolve on that short a time-scale. In nuclear physics, a nanosecond is a very long time interval—almost an eternity. But for us, it is short beyond comprehension. Why is that?

One could argue that our "clock speed" is limited by biological processes that are irreversible in the sense of the second law of thermodynamics. The neurons in our brain send signals to other neurons via traveling voltage pulses (called action potentials). These voltage signals propagate due to the diffusion of ions across the cell membrane in response to various concentration gradients. At a synapse, neurotransmitters get released and travel by molecular diffusion across a gap to reach the dendrite of the next neuron, where it can bind to receptor sites. On a more basic level, our metabolism converts low-entropy into higher-entropy stuff. We breathe in oxygen so that our body can burn hydrocarbon-chains into CO_2. In the process, energy is generated part of which is transferred into our environment as heat by means of conduction, evaporation and radiation. All of these processes necessary for human life are irreversible and increase the overall entropy of the universe. And, importantly, they all proceed at their own characteristic speed. It is not surprising that we cannot possibly register anything faster than the time it takes for neurotransmitters, say, to diffuse into the gap between neurons. This time-scale is described by Boltzmann's non-equilibrium statistical mechanics, but for us it has a much more personal meaning—it sets the smallest unit of perceivable time.

The fact that cellular processes (processes like ion diffusion that are thermodynamically irreversible) not only form the basis for our perception of time, but are involved in manufacturing a direction of time, should give us a moment of pause. A very strong case can be made that the very directionality of time is an emergent property—a byproduct, if you will, of the aggregation of particles.

Nobel Laureate chemist Ilya Prigogine has gone a step further and compellingly argued that the arrow of time is also intricately connected to, and indeed constructed by, the rise of *dissipative structures* in complex systems far from thermodynamic equilibrium [41]. Such coherent structures typically emerge from turbulence and instability. Hence, far from enabling predictability and determinism, Prigogine sees the arrow of time as arising precisely because our world is fundamentally non-deterministic. We will return to the subject of chaos and order later in the book. First, however, let's investigate other basic features of our macro-world (besides time) that similarly emerge when microscopic particles aggregate and arrange themselves into macroscopic matter.

3.2   A Taste of Condensed Matter

Condensed-matter physics is still the largest branch of physics (in terms of funding and numbers of students), but it may not be very prominent in the public eye, or perceived as particularly 'sexy'. Much flashier are astrophysics with its focus on the very large—stars and extrasolar planets, galaxies and dark matter, the universe and cosmology—and elementary particle physics with its focus on the very small—leptons and hadrons, quarks and gluons, supersymmetry and strings. Condensed-matter physics is somewhere in between, only vaguely defined for most people, somehow related to material science and chemistry perhaps. In reality, though, the field's reach has been pretty sweeping, and within physics it has contributed in fundamental ways to our understanding of nature. From time to time it has generated enormous excitement that goes well beyond its own borders.

I remember having the same sort of blank feeling about condensed matter physics in college as most people. As long as I can remember, I had liked physics and astronomy, so it was not surprising to anyone when I decided to major in the subject in college. But physics is an old science, and it takes a good while to get to the cutting edge. So by my junior year, I had not advanced enough to really have any exposure to the contemporary branches of physics, and I started developing the urge to at least get a taste of an actual research field of physics. Sure, I was enjoying the courses I was taking at Denison University—a small liberal-arts college in Granville, Ohio—and I really liked the way my professors made things interesting and approachable. In fact, I adored many of my physics professors and would later do research under their guidance in my senior year. At this point, I had already taken first courses in quantum physics, electrodynamics, and thermodynamics. A lot of what I was learning formed the foundations of contemporary physics, really cool stuff, but it was not contemporary physics itself. You couldn't go out and get a Ph.D. in quantum mechanics per se.

At the same time, however, I remember being a bit insecure. Did I really have enough of the fundamentals under my belt to do this?

Research papers in the leading journals were still hardly penetrable, research colloquia were usually only comprehensible for the first 10–15 min. And I hadn't yet had the chance to take some of the upper-level courses at Denison. Nevertheless, I decided I needed to take the risk and get some exposure. I needed to clarify for myself if I would actually like physics as it was practiced now, not as it was presented in textbooks. So I decided to apply to an internship program that Denison had arranged with Oak Ridge National Laboratories.

Oak Ridge was a famous place to physicists. It had played a big role in the Manhattan project where it was second only to Los Alamos. It was in Oak Ridge that the Uranium was enriched so that it could become useful in a bomb. It was Oak Ridge that built the first useful nuclear reactor, cryptically called the X-10. After the war, Oak Ridge had naturally remained a major center for nuclear physics, and started really spearheading nuclear reactor technology to be used not just for the production of electrical energy but also for propulsion of submarines, for instance. In the late 1960s and 1970s, Oak Ridge with its historical expertise in nuclear physics became a major center for fusion research. However, it soon became clear that the prospects of a fusion reactor were not imminent. At the same time, public awareness of the risks of nuclear energy began to grow and then really surged after the Three-Mile-Island accident in 1979. It was clear that Oak Ridge needed to broaden its scientific scope and diversify its mission. It turns out that the stream of neutrons, produced by the research reactors, can be used to probe material structure, and so it seemed natural for Oak Ridge to shift emphasis to material science, which also held the promise of industrial payoffs.

By the time I got there in 1995, all kinds of research was going on within its sprawling (but fenced off) premises, spanning all of the sciences. To me, the place was absolutely huge. I had requested to be part of the solid-state division and had been accepted. Before leaving campus and driving to Tennessee in my tiny old Dodge Champ, I had called up the head of the division, Dr. Lynn Boatner. "Is there anything I should read up on before getting started?", I asked hesitantly. "You're going to be fine. Just get here in one piece", was the reply. "I haven't had any solid state course, yet", I confided, thinking I should fess up to this gap in my education. "That'll be fine. Nothing to worry about, son." I remember being relieved but also a bit puzzled when I had finished the phone call. Why wasn't it going to be important for me to know some solid-state theory beforehand?

Within the first couple of days of starting the semester-long internship the answer became clear to me. I was assigned to work under a postdoc from Switzerland and was immediately thrust into the ongoing project. It had to do with growing crystals of a perovskite structure and changing its lattice spacing by introducing impurity atoms. So I was soon mixing chemicals in the stoichometrically correct proportions, baking them into a melt and then cooling them slowly down to crystallize. I learned how to operate crystal cutting machines, and the polishing of the crystals. I got acquainted with the X-ray powder diffractometer and the Laue back-scattering machine they had there, and was soon checking X-ray diffraction peaks of various samples. At first I was under close supervision by my postdoc, and Dr. Boatner himself would sometime look over my shoulders as well, but as the semester wore on they started trusting me more and more to do certain things by myself.

Lynn Boatner was an interesting man. He kind of fit my image of the old-school, no-nonsense scientist; always semi-formally dressed, very dedicated and hard-working, someone you wouldn't want to cross, but also jovial and good at telling fun stories about science personalities that had come through the lab, historical tidbits, or the curious mishaps that had happened in the lab. He was the head of the division at the time, but he seemed to be closely involved in a lot of the really routine, practical stuff too. He knew and was personally close with some of the technicians in the division. And he was really into crystals and the art of crystal growth. I remember him proudly showing off some massive single crystals that the team had managed to grow.

One day he came around to show us all something special. It was a single crystal the size of a pebble, but the inside of it was deep blue. As you went out from the center, however, the crystal lost more and more of its blueness and actually became entirely clear at the borders. It was pretty neat to see. I asked how that was possible? It turned out to be an impurity gradient—lots of impurities at the center that gave the crystal its blue color, and then less and less impurity concentration farther out. Unfortunately, I never learned how they had managed to grow this crystal.

The origin of color in minerals is actually interesting science in itself. Alkali-halide crystals, such as sodium chloride or potassium chloride, are naturally clear. Electrons cannot absorb photons of visible light in these lattices due to the particulars of the electronic band structure. Oftentimes, however, those crystals are found to be colored. So where does the color come from? The quick answer is lattice defects or chemical impurities. An example of the former is the so-called F-center, where F stands for the German *Farbe* meaning color. It is simply a negative ion vacancy—a chlorine ion is missing at a vertex in the lattice usually reserved for it. This means that locally there is now an unbalanced positive charge which, not surprisingly, attracts an electron into it.

It is this electron trapped in the vacancy site that can absorb photons of particular colors. One can think of this electron bound to the vacancy site as forming its own "atom." The only difference is that the electron is held in place by more complicated *crystal fields* instead of by a single proton. But again, the electron is not to be thought of as a point particle; rather, its wavefunction spreads out over all the available space, being somewhat concentrated at the sites of the positive ions surrounding the vacancy. Just like ordinary atoms, these F-centers also have excited states, and so photons of a particular wavelength can be absorbed. Which colors get absorbed depends principally on the lattice. In sodium chloride, F-centers produce a yellow tint, and in potassium chloride, they turn the crystals magenta [42]. But it also depends on whether other such F-center are nearby in the crystal and in what spatial configuration.

Other minerals get their color via trace amounts of impurities, which are ions absent in the pure form of the crystal. So, for instance, diamond in its pure form is clear, but when a relatively small number of

Boron atoms is added into its carbon-lattice, diamond turns blue. When nitrogen is included in trace amounts, diamond turns yellow. Of course, there is nothing yellow about nitrogen atoms or molecules themselves—nitrogen gas is colorless. Only through the interplay between the ion and the host lattice does the color arise. This point is further illustrated by the example of chromium impurities. When small amounts (typically less than 0.1 percent) of chromium, Cr^{3+}, replace the native aluminum ions within the aluminum oxide lattice, we get Ruby—a vibrantly red gem-stone. However, chromium replacing aluminum in the Emerald gem, gives us a green hue. The same impurity ion replacing the same host ion yields different colors—red in Ruby and green in Emerald. The color is not contained in the ion itself, whether impurity or host, but arises through the interaction between ion and host lattice (Fig. 3.2).

It was these kinds of conversations, experiences and interactions at Oak Ridge got me excited about the subject—I mean who wouldn't be turned on by crystals and gemstones, and things like the origin of color! By the end of that semester, I was pretty sure I was going to like graduate school and condensed-matter physics. I had been

Fig. 3.2 Ruby and emerald: the two colors are due to the same impurity ion, chromium, replacing the same native ion, aluminum, in the two host lattices

given a crash course in material science, and most importantly, I had learned something about myself: even though I had gravitated to physics initially because I liked the mathematical language of it, I also really enjoyed doing the hands-on parts, the nitty-gritty jobs—the growing, cutting and polishing of crystals, the programming of the furnaces, operating the X-ray equipment. It was all fun! I had been exposed to real down-to-earth material science at Oak Ridge, and I found myself liking it. A formal introduction into the larger subject of condensed-matter physics would have to await graduate school, but I started to feel that I was ready for it.

So what defines this larger field of condensed-matter physics; what makes it distinctive? In a nutshell, it aims to explain the states of matter we see from physical law. Whether we are talking about the familiar crystalline solids and liquids, or classes of materials like ferromagnets (permanent magnets) or semiconductors, or whether it be glasses and the less familiar quasi-crystals, or the even more exotic superfluids and superconductors, the objective is the same: to understand the measurable properties of these states of matter from basic physical principles. Its goal, therefore, is a quantum-mechanical and statistical theory of solids and liquids. In its modern tool-bag, so to speak, are techniques from quantum-field theory with its host of quantum excitations called quasi-particles. These quasi-particles go by such illustrious names as phonons, magnons, excitons, polarons, polaritons, Cooper pairs, and rotons—after a while you get the dizzying sense that you have joined the screenwriters for Star Trek, there is so much technical jargon involved. Nevertheless, these terms all refer to some process by which a solid (or another macroscopic phase of matter) can be excited, the ways in which it can store energy. These discrete excitations are also sometimes called the collective "degrees of freedom."

Why would the word *freedom* be used in this context? The way it make sense to me is this: let's say I deposited some energy in the solid, so energy comes in from the outside. But I cannot dictate to the solid in what form that energy ultimately has to manifest itself inside the solid. The solid, in a way, has a number of "choices" to make as to where to put that energy. So, for instance, it could put it into the vibration of its lattice, or it could give the energy to its electrons, or if it is a magnetic solid, it could decide to flip a spin the other way. There are many choices, hence the term degrees of freedom.

Now it turns out that these various choices all have one thing in common: they don't just accept any old energy, they only accept discrete bundles of energy or chunks. The size of the chunks varies, but there is always a smallest indivisible unit. And these chunks are what we call quasi-particles. These 'newly minted' particles sound esoteric, but in fact they are considered as representing an aspect of reality; they are, in a sense, as real as any construct or model in physics. Being absolutely indispensable for a correct and reliable description of material properties, they might be viewed as an embodiment of *functional emergence*.

Take, for instance, the magnon. The classical model of the magnon is a spin-wave, just as the classical model of the photon is the electromagnetic wave. Both transcend their classical description by representing a smallest quantum of energy. The photon is the smallest chunk of energy in an electromagnetic wave of a certain wavelength, and the magnon is the smallest chunk of energy in a spin-wave of a certain wavelength. What is a spin-wave? It is wave motion (in the form of wobbles like those of a spinning top) in the orientations of aligned spins of a ferromagnet. An illustration of a spin wave is shown in Fig. 3.3. We see that the spins, depicted by arrows, are not all perfectly aligned as they would be if they had not been disturbed. No, some energy was deposited in the spins and now they are seen to tilt away by some angle and form a wave-like pattern; if we could watch this wave over time, we would see it propagate in the direction of the arrow on the side.

Something similar is true for the electromagnetic wave, also depicted in the figure. Now the little arrows represent the electric and magnetic fields that exist at each point in space. And while the pictures look pretty similar, there may appear to be a fundamental difference between the two. The spin-wave is contingent upon there being a magnetic lattice first. The arrows refer to physical spins. The electromagnetic wave

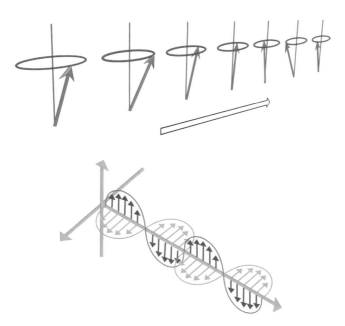

Fig. 3.3 *Top*: A group of spins excited by a spin-wave. They are no longer pointing exactly in one and the same direction but rather their tips trace out small circles. *Bottom*: Compare this two the classical representation of an electromagnetic wave, also known as light. Here the *arrows* refer to the electric and magnetic vector field. (Image from: missionscience.nasa.gov/ems)

travels through vacuum and doesn't need any medium within which to exist. The arrows refer to fields. It is easy, however, to exaggerate this distinction. From a modern viewpoint, the vacuum itself is not nothingness, but rather a canvas teaming with matter and anti-matter, continuously drifting in and out of existence—it is a kind of medium.

In that both the photon and the magnons are particles of the kind postulated by quantum field theory, they share other common features. Although they are given the name particle because they hold a fixed amount of energy, they are generally not spatially localized, or occupy a small region of space, but instead they are spread out over large regions of the available space. So the pictures in Fig. 3.3 are not too far off. However, these waves can only accept certain amounts of energy; their energy is quantized! The associated particles are packages of an extended quantum wave. Only sometimes can we contrive things to make them localize in space.

In fact, that was roughly the topic of my dissertation: how to spatially localize magnons and how to keep them localized over time by exploiting nonlinearities in the underlying lattice. One of the curious things I found at the time was that the physical shape of the magnetic crystal (whether it was a plate or a rod, for instance) was an important factor in the localization of magnons [43]. How come? That wouldn't really make sense if magnons were microscopic particles occupying tiny regions of space. How could they then possibly know about the macroscopic shape of the crystal, about boundaries that are hundreds of million lattice sites away? No, instead a magnon is a wave that spreads out over the entire crystal and can thus sense its boundaries.

In a way, condensed-matter physics is the ideal branch of physics (and ideal field of science more broadly) to contribute very deeply to a discussion on reduction and emergence. Whereas elementary particle physics is concerned only with the "parts," meaning with the smallest constituents of matter, condensed-matter physics wrestles with both "parts" and "whole," with both atoms and their aggregation into macroscopic matter. It looks in detail at phase transitions where matter reorganizes itself on the macro-scale and in the process redefines the relations between the constituent parts and often even their essential nature. In its very outlook and focus, it is ideally suited to give us insight into the mutual relationship between emergent properties of matter, the theoretical description at that level, but also the microscopic makeup of matter.

Furthermore, unlike biology or neuroscience, where the complexity is again exponentially higher, condensed-matter physics treats phenomena just complex enough to exhibit profound instances of emergent behavior, but also not so far removed from basic physical law that a mathematical description is still possible. Whereas speculations about reduction and emergence in neuroscience, for instance, still have a somewhat qualitative feel, condensed-matter physics provides us with a more quantitative route to exploring emergence. So we have found a good next topic to deepen our discussion on emergent behavior. Let's delve right in and start with a few basic examples that speak to these larger issues.

3.3 Macroscopic Phases

Graphite and diamond are both crystalline solids made up entirely of carbon atoms. Nothing else, just carbon. Yet, this is the only thing they share in common. Graphite is a black substance that absorbs visible light of any color. It cleaves easily; when you rub it against another surface parts of it tend to come off—hence its usefulness in pencils. It conducts electricity reasonably well, but not too well—it is a semi-metal in the scientific lingo, often used in electrical resistors. It is comparatively light in weight, and quite inexpensive. Diamond on the other hand is completely clear; it does not absorb visible light of any color. It is the hardest solid there is—hence its use in diamond-impregnated blades to cut metal and glass. It is an electrical insulator and most importantly, it is also very expensive, as many of us know from personal experience. Macroscopically, the two share nothing in common. Anything you could want to measure about solids, the results will be different for the two. Whether it be thermal, optical, elastic, mechanical, acoustical, or electrical properties, there is really no similarity between graphite and diamond. Yet both are made up entirely of carbon atoms, and if we burned a piece of graphite and a piece of diamond, the CO_2 that would be produced would be completely indistinguishable.

What causes these big differences then? Not surprisingly, it has to do with the spatial arrangement of the carbon atoms in the lattice, also known as the lattice structure. In diamond, the atoms are tetrahedrally bonded to form a lattice, a geometric twist on the cubic arrangement. For graphite, on the other hand, we get a lattice of a very different symmetry: layer stacked upon layers of honeycomb carbon sheets, as shown in the figure. The two atomic arrangements are quite different even on a visual level. It turns out that it is this arrangement, the

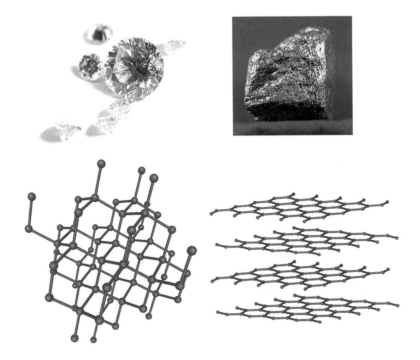

Fig. 3.4 The crystal structure of diamond and graphite. Both are made entirely of carbon atoms, yet their macroscopic properties are completely different. (Image by Itub, https://commons.wikimedia.org/wiki/File:Diamond_and_graphite.jpg, licensed under CC BY-SA 3.0)

organizational principle of the lattice, if you will, that determines all that we care about with respect to these solids. Next time you buy a diamond, remember that you are really paying for the tetrahedral symmetry of the underlying lattice, not the carbon (Fig. 3.4).

Of course, the same is true for all solids. Structure matters! You may be wondering whether this observation is trivial or profound; even upon further contemplation it can sometimes appear as either or both. However, the main lesson to take away for right now is that macroscopic properties of matter cannot be understood solely by the constituent parts in isolation (the atoms or molecules) and is not somehow already contained in these. If the same parts can give rise to altogether different objects depending upon how they are assembled together, then the properties of the object cannot already be contained within the parts. As Philip Anderson recently put it [44],

> It should be a little mysterious to you that some of the simplest properties of everyday materials have absolutely no connection to the properties of the atoms of which they are made, or of the very general laws which rule the atom. In no sense is an atom of copper red, shiny, ductile, and cool to the touch [...]. When we put a lot of atoms of copper together they become a metal, something conceptually new, that wasn't contained in the separate atoms themselves.

Another instructive example with which we have daily familiarity is water. Could the properties of water be deduced from an intricate knowledge of the single water molecule? Could we derive the phases of water, their properties and boundaries, from *first principles*, meaning from a molecular theory? These were the questions I still remember Neil Ashcroft presenting to a group of us. "Could an alien who had never seen or experienced liquid water before nevertheless derive its macroscopic properties and therefore not be surprised upon experiencing it on earth?" is how he roughly phrased it.

At the time, I had almost finished my undergraduate degree in physics and was visiting some of the grad schools to which I had applied. This brought me to Cornell University, my top choice already going in, but one that had not offered me a fellowship but "just" a teaching assistantship. My undergraduate advisers had stressed the importance of actually visiting the programs to which I had been accepted. So here I was with a group of other seniors, all of us in the same boat, at the top floor of Clark Hall overlooking beautiful Ithaca and lake Cayuga, where the physics department was housed. And there was the famous Neil Ashcroft, distinguished professor of physics and first author of the definitive text on solid-state physics, directing this slightly unusual question our way.

It wasn't immediately clear to me where he was going with it. As I suspect many of my peers in the room, I had not had a formal solid-state physics course in college, and so my thinking was dominated by experiences in the core subject areas—classical mechanics, electrodynamics, quantum mechanics, in other words the stuff that forms the typical staple of the undergraduate physics major. And in these subjects, the problems were by and large carefully selected for their analytical solvability. The problems I had encountered up to that point had solutions. Yes, some mathematical sophistication, like multi-variable calculus, was often needed to solve them, but they were doable with enough dedication. And yes, sometimes simplifying assumption would have to be made, or certain approximation techniques employed. But in the end, we would get some analytical answer that was close enough to what would truly happen in nature. The hydrogen atom problem was solvable using the Schroedinger equation, even if it took up many pages to do so; the helium atom was not and no easy approximation seemed adequate, which explains why I had not encountered it in any course. In short, and perhaps not surprisingly, my inclination was to answer Ashcroft's question in the affirmative, although I was (luckily) too self-conscious to actually volunteer this opinion.

So I was intrigued when he offered a different view. He granted that one could argue that there were a few clues about H_2O that an alien could use to his advantage in this quest. For one, it is a very polar molecule, with the oxygen drawing electrons to it and becoming somewhat negatively charged, leaving the hydrogen atoms somewhat electrically positive. In this way, each molecule acts like a little electric dipole with a positive end and a negative end, as shown in the figure (Fig. 3.5). We know from classical electromagnetic theory that two such electric dipoles exert forces on one another via the electric fields they produce; they can attract. In this way, the alien could predict that water is not easily modeled as an ideal gas of non-interacting particles (unless the temperature is very high), and that it would have a tendency to condense into a liquid at a temperature that is not too low, and actually very high compared to, say, helium. But even this presupposes that the alien had seen liquids (other than water) before, otherwise the term liquid would have little meaning for him. So, if the alien had come from Mars, where no liquid water, or other liquids (with the exception of mercury), can exist due to the low ambient pressure, the alien would have no concept whatsoever. If a "gaseous" alien had arrived from

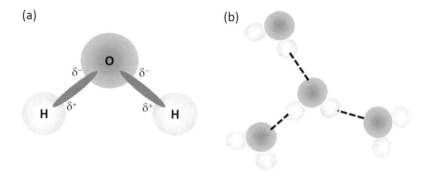

Fig. 3.5 (**a**) A single water molecule. The oxygen is slightly electro-negative, whereas the hydrogens are electro-positive. (**b**) Water molecules attract one other by virtue of each representing a little electrical dipole, with an electrically positive and a negative end. The weak attraction is called a hydrogen bond

Jupiter, then even the solid phase would seem like the most wondrously incomprehensible thing to him.

As I remember it, Ashcroft seemed to indicate that we might not get much further than this in our quest. While there is no doubt that knowledge of the individual molecule does yield some general information of this sort, it might not be able to do much more. Even with the previous example of diamond and graphite, the two different possible lattice structures of carbon might have been anticipated based on detailed knowledge of the electronic states (and their hybridization) of the carbon atom. It should be pointed out, however, that even this observation is not rigorously and exactly derivable from a theory starting at the lowest level with six mutually interacting electrons. Let's return to water, however, as things get a bit more complicated in the solid state, and a glimpse of the full complexity is useful here in order to appreciate the larger point.

It turns out that water-ice has multiple distinct phases—all distinct in lattice structure—that manifest themselves in different regions of pressure and temperature. They are shown in Fig. 3.6. At latest count, there are 16 different crystalline phases, ranging from hexagonal, to cubic, to tetragonal, and rhombohedral lattices. None of the discovered phases of water-ice were predicted from first principles [5]. (Recently,

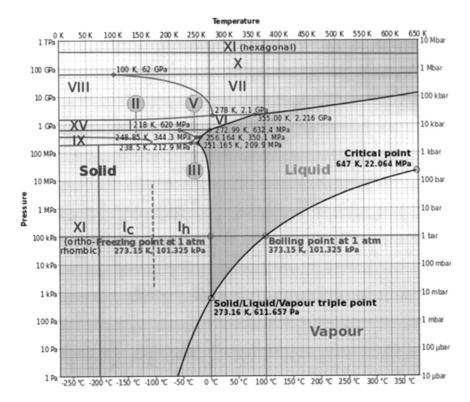

Fig. 3.6 The different phases of water displayed in the standard pressure-temperature diagram. In addition to the three known phases of water, there are other solid phases less well known that "live" either at higher pressures or lower temperatures. Their crystalline structure and macroscopic properties can differ substantially from the regular ice. (Image by Cmglee, https://commons.wikimedia.org/wiki/File:Phase_diagram_of_water.svg, licensed under CC BY-SA 3.0)

some progress has been made on the computer modeling front, but even this approach does not really classify as "first principles" research, but makes use of powerful computing resources to perform numerical experiments.) On top of these crystalline phases, there are also ways of producing amorphous water ice, a disordered solid called a glass, described by an absence of long-range order in favor of only short-range order. There are three distinct amorphous phases that have been discovered.

Each of these solid phases, of course, has its own set of physical/material properties. Things like density change abruptly from one to the next, but also things like heat conductivity, compressibility, entropy or tensile strength, and so on. We all know that ice is lighter than liquid water, but that is only true for the usual hexagonal ice, and not for the other ices that live at higher pressures. For those, it turns out, the density jumps up when liquid water freezes. Ice-cubes made of these ices would therefore not float in the liquid water but sink. The upshot is that even just in the solid state, water is extremely complex. When water molecules aggregate into macroscopic phases, many possibilities emerge in which the molecules can interact and arrange themselves. It is hard to see how this complexity could be accounted for in the simplicity of the single water-molecule.

We don't have to stop there. Let's take a look at the liquid phase for a moment. An important and measurable attribute of a liquid is its viscosity, that is how easily it flows or loosely speaking how 'gooey' it is. Would we really expect this viscosity of liquid water to be derivable from atomic and molecular theory? Viscosity is not something that even makes sense at the level of the single molecule or even at the level of hundreds of molecules. It is essentially a macroscopic phenomenon that arises in the limit of very large numbers of molecules interacting with one another.

The line in a phase diagram (of pressure versus temperature) separating liquid water from water vapor is called the phase boundary, or coexistence line, but it terminates at high pressures and temperatures in a critical point beyond which the liquid and gas phases become indistinguishable (see Fig. 3.6). If we follow the phase boundary line in the opposite direction to lower pressures and temperature we arrive at another significant point, namely the triple point where three phases (gas, liquid, solid) meet. Would we really expect that the location of these points could be predicted by a molecular theory? If so, it certainly hasn't happened yet, and isn't likely to happen. We run into "walls of complexity" [5] that are for all practical purposes not scalable.

Looking back, I suspect that Prof. Neil Ashcroft knew that from our undergraduate studies we had implicitly internalized reductionism to some extent, and he wanted to challenge us and stretch our imagination a bit. Were we as naive as to think that all problems in physics could be solved from first principles? If so, we were in for a rude awakening in graduate school, he might have been thinking.

3.4   Helium-3

I decided to enroll in the physics Ph.D. program at Cornell University and started in the fall semester of 1996. I found an off-campus basement apartment, and everyday I would walk or bike up the steep slopes of Buffalo Street, across Cascadilla gorge and further up the campus hills, taking in the natural beauty of the place along the way; I crossed the main quad of the university with the statues of the two founders, Ezra Cornell and Andrew White, facing each other on either end, until I finally arrived at Clark Hall. It was an exciting time in my life. Cornell was a new environment, where I was surrounded all day by very smart and intense people with diverse sets of ideas. We were an inquisitive and ambitious bunch, but not overly competitive with each other; most of us would study together and help out if someone got badly stuck on problem sets, and we would do fun stuff together in our spare time. Taking graduate courses was challenging, and the level of difficulty was something I definitely struggled with at first, but the amount of stuff I was absorbing, the new physics I was learning—it also felt exhilarating. And the intellectual freedom and the absence of close supervision felt liberating. There was a sense of individual possibility, but I also felt like I was part of something bigger.

One morning that first semester, when I was coming into Clark Hall, it must have been sometime in October, there was palpable excitement in the corridors of the physics building. People seemed to be walking with an extra bounce in their step, the general mood was strangely festive. Someone came running up to me, shouting "Did you hear, David Lee has won?" My curiosity definitely peaked. I had known Professor David

Lee, primarily because he was a teaching assistant for a different section of the same introductory physics course for engineers as I. At the time, I had found this very endearing but also a bit unusual. All the other TAs were new graduate students paid to lead the recitation sections of these large intro-physics courses. That's how most of us earned enough to pay our rents. But then there was David Lee, who as a full professor seemed a bit out-of-place in the TA meetings, and the marathon grading-sessions that we periodically had throughout the semester. And, most importantly, David Lee was approachable, folksy, down-to-earth, not as "professorial" as most other professors in the department. "Won what?" I asked. "He just won the Nobel Prize, man, where have you been" was the response. Wow! I had no idea that our Prof. David Lee, the one that we TA'ed with, was even considered or had done anything so path-breaking; he was that low-key and humble of a person.

It turned out that two Cornell professors, Robert Richardson and David Lee, had won the Nobel Prize in physics that year together with their former student Douglas Osheroff, for their discovery of superfluidity in helium-3. They had found this effect by serendipity. Serendipity in science is different from mere luck; it's more accurate to describe it as that extra little bit of good fortune that comes to those that have tried, come up short and tried again, persistently so, but who also kept an open mind throughout the process. In this case, Lee and Richardson had spent years building up their low-temperature physics lab in the basement of Clark Hall.

They and others in the Cornell low-temperature group had these amazingly impressive laboratories of unsurpassed experimental sophistication—the kind that you would take a visitor to see if you really wanted to impress them. There were massive cryostats, the containers used to cool things down to low temperatures, literally hanging from the ceiling. There were dewars, vacuum pumps, various pump lines and evacuated fill lines, electronic instrumentation to control currents in magnets, instruments to read pressure and track resistances of thermometers. Sometimes these low-temperature setups were combined with NMR and microwave measurements and apparatus.

I always had a mix of envy and pity for the graduate student brave enough to join these groups; it was a little-kept secret that it would likely take some of them 10 years to get their Ph.D.s—there was so much instrumentation, so many experimental techniques to master, before one could even get started on one's own project. Osheroff was one of those grad students in 1971 (although he would do it in 6 years). At that time, with their ability to produce ultra-low temperature, Lee, Richardson and Osheroff wanted to look at the rare isotope of helium, helium-3, with its two protons but only one neutron in the nucleus. It was, of course, well known that regular helium—helium-4—underwent a transition below 2.17 K, about two degrees above absolute zero, and became a superfluid—a kind of macroscopic quantum state of matter. But it was not clear what would happen to helium-3 at low temperatures. Nothing happened to it just below 2 K to be sure—it remained a regular liquid with fairly ordinary properties.

That this must be so was again not surprising to physicists. All particles belong to one of two broad classes—*fermions* or *bosons*. The basic distinction between these two classes of particles is how they act in the presence of other copies of themselves. Loosely speaking, fermions want to keep their distance from each other, whereas bosons like to hang around one another. Specifically, the so-called *Pauli exclusion principle* in quantum mechanics says that no two (or more) fermions can occupy the same quantum state, and as a consequence, fermions cannot condense into a single collective quantum state, as they would necessarily have to do in order to form a superfluid. It so turns out that the helium-4 atom is a boson, whereas helium-3, by virtue of having one less neutron, falls into the fermion camp.

So what happens to liquid helium-3 at very low temperatures? Would it remain a fairly ordinary liquid all the way down to absolute zero, or would it perhaps still find a way to turn itself into a superfluid? This was an unanswered question at the time, and the solution seemed so daunting that the Cornell team actually investigated a different question that they hoped would yield more easily to scrutiny. They decided to look at the helium-3 ice that only exists at high pressures. Would it turn itself into a magnet at low enough temperatures? By hard work and serendipity, they instead wound up answering the first question and in the process found one of the most intriguing state of matter anywhere in the universe.

Most ordinary substances do freeze and turn into a solid at low enough temperatures. In this phase, of course, atoms can be thought of as occupying regular lattice sites and vibrating about those positions with amplitudes small compared to the lattice spacing. At higher temperature this lattice vibration is more vigorous, and eventually gets too vigorous and the lattice melts. Upon cooling the crystal, the vibration must slow down, but the motion never completely stops but must continue even down to a temperature of absolute zero. Quantum mechanics demands a residual vibration at absolute zero, for the uncertainty principle says that no particle can have both well-defined position and velocity. If the atoms in a lattice stopped vibrating altogether, then we would know their velocity (zero) but also their position (the particular lattice vertex).

Thus, even at the lowest possible temperatures, there is a zero-point quantum motion left. How vigorous that motion is depends, not surprisingly, on the mass of the atom, with lighter masses leading to more vibrational amplitude. It also depends on the binding force that holds atoms in the lattice. What is interesting about helium is that it is light, second only to hydrogen, and that its inter-atomic forces are very weak. Helium, after all, is an inert gas, so it doesn't interact much with anything else. These two factors combine to make helium the only substance that never freezes at ordinary pressures. It just can't coalesce into a solid, as the zero-point quantum motion of the helium atoms would be too large to sustain and would destroy the lattice. So, ordinary helium-4 stays liquid all the way to absolute zero, turning into a superfluid at 2.17 K, but staying liquid nonetheless. Only applying large pressures can get helium to solidify.

So what about helium-3? What happens to this rare isotope when cooled? The history is in fact intriguing in its own right, but we have to pick up the thread just a little earlier in time. Long before helium-3 became of interest, what had excited the imagination of many scientists was the question of whether ordinary gases could be liquified. By 1844 the great Michael Faraday had tried his hand at cooling down ordinary gases in order to turn as many of them into liquids and solids as he could. Being one of the finest experimentalists of his day, he had met success with almost all of them.

One of these gases was carbon dioxide, which at atmospheric pressures skips the liquid phase altogether and turns itself into a solid at $-109\,°F$. While improving upon earlier experiments by Thilorier on solid CO_2, or dry ice, Faraday observed that it "appears as a clear, transparent, crystalline, colourless body, like ice; so clear, indeed, that at times it was doubtful to the eye whether anything was in the tube." [45] He came extremely close to the modern accepted values for the triple point of CO_2, where all three of its phases meet, stating that the solid "melts at the temperature of $-70°$ or $-72°$ Fahr., and the solid carbonic acid is heavier than the fluid bathing it. The solid or liquid carbonic acid at this temperature has a pressure of 5.33 atmospheres nearly." Here the older term "carbonic acid" refers to carbon dioxide. Furthermore, Faraday noted that when he let some of the dry ice evaporate, it cooled even further, and that the lowest temperature that could be reached this way was $-148\,°F$.

Michael Faraday realized he could use this low temperature to see if other gases liquified—ones that had resisted it up to that point. Oxygen, nitrogen, and hydrogen were such gases, and so he now tried cooling them while simultaneously applying large pressures to them. Could he, in fact, make air into a liquid for the first time in human history? He went as high in pressure as 50 atmospheres for nitrogen, an enormous pressure that would make most vessels explode, but still no liquid. Nor did he have more luck with oxygen or hydrogen. Finally, he recorded in his notes that these gases "showed no signs of liquefaction when cooled by the carbonic acid bath in vacuo, even at the pressures expressed." As frigid as negative 148 °F sounds, it converts to 173 K and turns out to be rather hot compared to the temperatures at which oxygen or nitrogen become liquids: 90 K ($-298\,°F$) and 77 K ($-321\,°F$), respectively.

In retrospect, Faraday never stood a chance. It would take another 33 years, until 1877, to pull it off.

Hydrogen would resist liquification even longer. For a while, it seemed that perhaps this gas would never transform. Then, in 1898 the Scottish experimentalist James Dewar finally managed to liquify hydrogen at around 20 K, or −423 °F, and then even produced solid hydrogen at slightly lower temperatures. He had reached a scientific milestone, answered the long-standing hydrogen question, and basked in the glory at a special Royal Society meeting. He had a flair for the showy, and demonstrated his liquification for a live audience. Of all known gases only helium now remained; it was truly the last holdout. The race was on. Dutch physicist Heike Kamerlingh Onnes had also competed for the hydrogen prize and come up short. In science, coming in second is like never having competed at all; it wins you no praises or awards. But Onnes didn't give up, but decided to raise the stakes. He expanded his already massive laboratory to industrial proportions at Leiden University in the Netherlands with the new goal of liquifying helium.

How do you cool anything down to a temperature below the coldest thing in existence? The coldest thing in existence at that time was liquid hydrogen at 20 K, which could be cooled a bit further by pumping on its vapor above the liquid. But at that temperature, helium was happily gaseous. Even colder temperatures needed to be reached, but how? The technique used was based on a throttling process that forced the gas through a constriction. In 1908, Kamerlingh Onnes was finally ready. There had been a number of pre-cooling stages, and now the throttling process was chugging away. There was suspension in the room; people were trying to see if drops would form. Everyone was crowding around the glass cryostat, leaning in a bit, and keeping their eyes fixed on the bottom plate. And sure enough, little drops of helium were spotted. The team had done it. Heike Kamerling Onnes had finally beat James Dewar to the finish line. It had been a fierce competition which had involved twists and turns, setbacks and tragedy, and which also features strong personalities [46]. For me, the story also serves as a strong reminder that science always is a human endeavor.

Kamerlingh Onnes then went on to utilize this liquid helium to cool down metals and see if their electrical conductivity actually continued to improve. Or would the conductivity actually go to zero because the electrons would stop moving altogether, as some had speculated; if electrons can't move, then there is no electrical current, the thinking went. Onnes quickly shot down that hypothesis. In fact, quite the opposite was true. The conductivity became infinite (and the resistance dropped to zero) for all intents and purposes. In this way, his team had inadvertently pulled off a second major feat; they had discovered superconductivity in mercury samples—an immensely consequential discovery that would await a scientific explanation for almost another 50 years [47].

So ordinary helium (helium-4) became a liquid at around 4 K. But what about helium-3? In 1948, helium-3 became peripherally wrapped up in the cold war between the U.S. and the Soviet Union, due to the fact that tritium (^{3}H), a rare and unstable isotope of hydrogen, was key to making a hydrogen bomb—a weapon even more deadly than nuclear warheads—and tritium decays into helium-3 (^{3}He) via beta decay. So in the process of stockpiling tritium for use in weapons, the by-product helium-3 had to be separated out to keep the tritium unadulterated, and the Los Alamos staff figured they might as well collect it instead of just releasing it into the atmosphere. A fellow named Ed Hammel, who worked at Los Alamos National Laboratories at the time, was able to put this extraneous helium-3 to good use by cooling the gas and liquifying it for the first time. So helium-3 also liquified. But what had been produced was an ordinary liquid, and certainly not a superfluid.

Fast forward another decade, and finally an explanation for superconductivity in metals arrived in the form of the monumental Bardeen, Cooper and Schrieffer (BCS) theory, named after the three theorists who developed it in 1957 (J. Bardeen, L.N. Cooper, J.R. Schrieffer were awarded the 1972 Nobel Prize in Physics for this work). Their theory predicted a transmutation of electrons (fermions) into electron pairs (bosons) by means of an attractive interaction involving lattice vibrations. The reason this discovery is relevant to our brief history of helium-3 here is because only 2 years later, Philip Anderson (whom we have already met) predicted that a similar effect of 'transmutation due to pairing' could happen in liquid helium-3, and he initially predicted the superfluid transition at around a tenth of a Kelvin (80 mK).

Things were moving rapidly now. After all, there was now a concrete prediction available—something to shoot for. With this encouragement, experimentalists stepped up their search for the superfluid. But the initial excitement soon gave way to frustration, as it became clear that the 80 mK had been a substantial overestimate. In the late 1960s, John Wheatley (at the University of Illinois and then UC San Diego) got down to a temperature of 2 mK—way below the theorized transition temperature, but still no sign of the superfluid appeared. People in the community started to lose hope that it would ever be discovered, or that it was even for real [48]. That is until Lee, Richardson, and Osheroff tried their hand at it in the early 1970s.

Douglas Osheroff was sitting there in the lab late at night next to the cryostat with his home-built Pomeranchuk cell inside it, as I imagine him having done many a night before. Nighttime was and is precious for an experimental Ph.D. student. It took all day to simply prepare the system and cool everything down, that oftentimes you had to do the actual measurements at night. Plus things were quiet in the building, no interruptions nor distractions, perfect for concentrated work. So, it was another late-nighter in the building, the corridors were eerily deserted yet well lit by the harsh fluorescent-panel lighting; hardly any soul was around. In the lab, the fans of the instruments were gently humming away and data was gradually coming in and being plotted on a strip-chart recorder (in those days).

Douglas's project was to look for magnetic phase transitions of solid helium-3, not superfluidity [48]. This turned out to fortuitous, as we will see. It was about 2 AM in the morning and Douglas was riding along the melting line of helium-3—the line separating solid and liquid at high pressures—by compressing more and more of the liquid into the solid. In an odd twist of nature (first predicted by Russian physicist Pomeranchuk), liquid helium-3 turns out to be more ordered than solid helium-3 below a temperature of about 0.3 K, which interestingly means that you can cool the solid-liquid mixture by pressurizing it; usually applying pressure means heating, but not here. So that's what Douglas did; he increased the pressure thereby reaching lower and lower temperatures. All the while, the pressure curve (as a function of time) was being printed out on graph paper in real time by the strip-chart recorder. All of a sudden the slope of that line appeared to change. There was a noticeable kink in the curve.

At first, Douglas may have chalked it off to instrumental or equipment error, and it certainly would have been easy to dismiss this kink as some kind of artifact and think no more of it. Here is where persistence comes in, though. When Douglas retook the data a few days later, the kink was still there and amazingly enough it occurred at exactly the same pressure, and thus temperature, as before. This could not be a coincidence! What's more, being on the lookout for anomalies in the pressure trace, he now saw a second blip at an even lower temperature. Soon it dawned on him and his advisers that they were on to something big. After some additional NMR measurements, the group was eventually able to attribute these kinks to phase transitions in the liquid and to the arrival of two separate superfluid states in helium-3. The temperatures at which these superfluid phases manifest was observed at 2.6 mK and 1.8 mK, or about two thousand of a degree above absolute zero.

So why had it not been seen by John Wheatley who had gone down to 2 mK? The reason was that Douglas Osheroff was operating at high pressures—they initially wanted to study solid helium after all. High pressures, it turns out, also raises the superfluid transition temperatures and so inadvertently allowed them to make this discovery. We now know that at atmospheric pressures, the transition doesn't happen until you get down to half that temperature, or about 1 mK. With a little serendipity, the three had managed to produce something in the basement of Clark Hall that likely never existed anywhere else in the universe.

Helium-3 superconductivity is pretty cool (no pun intended), but how does it help us answer broader questions? What does it have to do with emergence? In a nutshell, the important aspect that allows helium-3 to transition to a superfluid state is the pairing up of single helium atoms. Two helium atoms form a pair, so what's the big deal? Well, according to the strange laws of quantum mechanics, a new type of quantum particle is born, namely the helium pair (see Fig. 3.7). Why is this new particle born? Because the collection of such particles can achieve a lower energy. It's a striking example of how the nature of the parts change when the whole reorganizes itself to minimize energy:

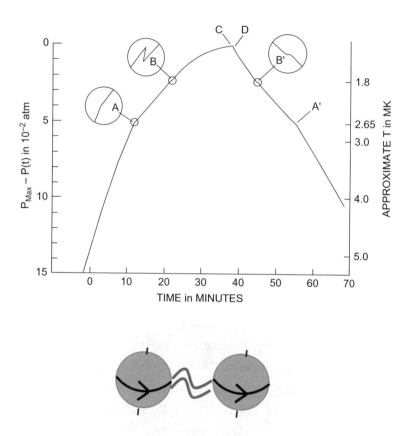

Fig. 3.7 (**a**) The kinks in the pressure curve observed by Douglas Osheroff that signaled the appearance of the superfluid phases in He-3. (The image appears in [49].) (**b**) Two helium-3 atoms pair up to form the analogy of a Cooper-pair in superconductivity

we get a macroscopically new state, the superfluid, but its parts have transformed. Instead of single helium atoms, the constituent parts are helium pairs—a new particle altogether.

In fact, we now understand both superfluid helium-3 and superconductors in much the same way. Individual electrons are fermions, but within superconductors they arrange themselves into pairs. "Wait, electrons form pairs?" you say, "I thought like charges repelled." Yes, normally they do, but in the context of the lattice this repulsion is much reduced due to electric shielding by the sea of background charges (other electrons and ions). And there is another mechanism at work that can actually cause attraction between electrons, a mechanism mediated by the lattice. Think of it this way: one electron somewhere in the lattice pulls the positive ions a bit closer in than they would normally be. The other electron at a distance thus sees a slight abundance of positive charge around the first electron, and feels attracted. Of course, the argument is symmetric—the second electron gets similarly attracted to the first. That's the basic idea. Now, this classical picture can only get us so far. Quantum mechanically, the explanation for the attractive interaction between this electron pair (of opposite momentum and spin) is that they exchange virtual *phonons*.

No, that's not a spelling mistake. A phonon is very analogous to the photon—the particle of light. It is the particle of sound, or atomic vibrations, in a crystal lattice. Within the crystal, sound is a wave very much like light in vacuum. There are only discrete energies that this sound wave can have; the energy is quantized into little chunks called phonons. Metaphysically, the two quantum particles are on an equal footing: the photon lives in the context of the vacuum and the phonon in the confines of the lattice, and mathematically there is not much difference either. In fact, the analogy goes even further. Photons mediate the electromagnetic interaction between charges; two electrons in mutual proximity exchange virtual photons and repel as a result. Here the term 'virtual' is used to indicate that the photon is emitted by one and quickly absorbed by the other electron before it could ever be observed by itself. The same thing now happens with the phonon.

Two lattice electrons can exchange phonons, but now the net result is attraction.

However strange this all sounds, the fact is that we have these electron pairs held together by phonons. The situation is rather like an entangled state of two electrons that we encountered in Chap. 2. The pair has different properties that are not reducible to the parts. The building blocks are single electrons which are fermions and as such obey the Pauli exclusion principle. The pair, however, is a boson and does not obey it. The pair interacts with another pair in a way that is fundamentally different from how four electrons would interact with one another. Four electrons would avoid each other like the plague even if we turned off the electrostatic repulsion (which we can in the lattice due to shielding). The Pauli exclusion principle forces electrons to avoid each other.

Two electron pairs, however, do the opposite—they like to be in the same state. The electron pair, more frequently called the Cooper pair, is not just some designation based on a mental grouping into pairs. No, it must be seen as a new physical entity with just as much claim to reality as the electron itself. In fact, in the superconductor, it's the Cooper pairs that form the basic building blocks, the particles responsible for the conduction of electrical current. It is these pairs that now coalesce into one collective quantum state in a process very much like Bose-Einstein condensation. Unlike the conduction of individual electrons, the pair moves around without any scattering or disturbance—hence the zero resistance of superconductors. The macroscopic consequence of Cooper pairs and their condensation is that if you set up a current in a superconducting ring once, that current never diminishes. It keeps going forever. Resistance, it turns out, is not only futile, but exactly zero.

A very analogous story also describes the formation of the superfluid in helium-3. Here the role of the electron is now played by the helium-3 atom which is (as we said earlier) also a fermion. By itself, therefore, helium-3 atoms cannot condense into a single quantum state. In fact, they individually have to do their own thing. If anything like collective condensation is to happen, they have to pair up. While the end result is very similar to superconductivity, the mechanism here is a bit different. It has to be different, since we have no lattice to work with here, and no lattice also means no phonons. The attractive interaction is actually intrinsic to the helium atoms. At close proximity, two helium atoms will repel very strongly, but at intermediate distances, there is in fact a small attractive force. Again, at low enough temperatures, this attractive force can sustain stable pairs of helium-3 atoms. Crucially, this transmutation of the single helium-3 to Cooper pair is not at all like putting two Lego pieces together. It is more reminiscent of entanglement. The pair transcends the parts and becomes fundamentally a new entity—a boson.

The question is: Why did the pairing occur in the first place? What made the atoms pair up? The answer must partly lie in the new collective quantum state that this regrouping enables. This microscopic pairing makes possible a macroscopically coherent quantum state in which the parts give up their individual character altogether. This is akin to the famous Bose-Einstein condensation that "fuses" together entities that formerly retained some individual reality. Now the parts have ceased to be parts but instead have fused into a state without any trace of distinguishability. A new collective quantum state has emerged that is irreducible to a simple collection of single pairs of particles and that has, in some deep sense, destroyed the discreteness and identity of its former constituent parts. In a sense, the reason for the arrangement into atomic pairs was to usher in this new and strange macroscopic state. The only way the new macro-state could emerge is because it redefined the nature of its own parts—an important hallmark of emergence.

Chapter 4
When Matter Reorganizes Itself

4.1 Self-Consistency

At Dickinson College, like at so many other colleges and universities, the hallways of the physics building are lined with framed posters put out by the American Physical Society. You may have seen them, too: they showcase, decade by decade, major personal achievements, important general advances in knowledge, and pivotal events in our discipline. Together, they represent something of an arrow of time for physics. Each poster starts with a written paragraph describing in broad strokes what happened in the respective decade, with a big picture of a prominent physicist from that era in the top corners. So each day I walk past the likes of Marie Curie, Albert Einstein, Paul Dirac, and Richard Feynman on my way to the office.

One day, within a framed box of the 1970s poster something caught my eye as I was walking by; there was a smaller picture that had something familiar about it. It was taken in 1982, and featured a smiling Kenneth Wilson in front of a long, sort of old-fashioned blackboard partially covered up with pieces of paper; on those pieces was written in large letters "Congratulations, Ken." Now, I remembered where I had seen something like that before. It was at the 1996 reception for Professors Lee and Richardson. And it even appeared to be the same room at the top of Clark Hall where Neil Ashcroft had graciously taken time to challenge the naive reductionism of myself and other prospective grad students. Pieces of paper covered up the blackboard congratulating the two 1996 winners in just the same way.

By the time I had arrived at Cornell as a grad student, Kenneth Wilson had left for Ohio State, but his name was very much present still within the physics department. He had received the Nobel prize for his theoretical work on phase transitions and, in particular, for his formulation of the "renormalization group" approach to the problem. It turned out to be an approach that would make a lasting impact on many branches of physics, including (interestingly enough) elementary particle physics. First and foremost, it settled an important and long-standing question: What in the world was happening at the critical point? That's the point where matter "decides" to reorganize itself on a macroscopic scale from one phase to another.

It was known experimentally that things sort of went haywire at that point. Heat capacities and other measures like the compressibility of a gas or the susceptibility of a ferromagnet went to infinity at that point, approaching a vertical asymptote from both sides. Infinite compressibility, what does that even mean? Basically it means that there is absolutely no push-back against squeezing the gas; none—the pressure does not budge. Similar story for magnetic susceptibility—turn on the tiniest magnetic field, and the magnet responds by developing a large magnetization. The slightest perturbations have huge effects when matter is at the critical point between two phases (separated by a second-order transition).

But there is one more important quantity that goes haywire. Let me start by way of example. Critical opalescence in sulfurhexafluride (SF_6) is a pretty neat demonstration that I show to my thermodynamics class every time I get to teach the course; it's one that really gets a reaction out of the students. What you have is some SF_6 trapped in a pressure cell. A pressure cell is a fancy name for a sealed container that can accommodate a high pressure on the inside. In the cell used here, you can actually see inside of it because the sides have windows. Won't the glass shatter at high pressures? Well, no worries, the cell uses pieces of glass that are particularly thick and held in place by a strong steel frame. So now imagine gradually filling this pressure cell with SF_6. In fact, you squeeze so much of that gas into it that the pressure starts getting quite appreciable; it would now undoubtedly shatter most ordinary containers. Still you keep going. The pressure now rises above the threshold for condensation and you see droplets of liquid SF_6 fall to the bottom. You continue filling until about half of the cell is occupied by fluid. Now you stop and seal off the fill hole of the cell. The bottom half is fluid, and the top half of the available space is, of course, not vacuum, but SF_6 gas.

A stable equilibrium between liquid and gas is formed—at the so-called vapor pressure. The separation of the liquid and the gas can be seen very clearly by shining light through the cell and projecting the image on a screen—an overhead projector (laid on its side) works pretty well for this purpose. The projected image shows a round circle (the pressure cell), and cutting it in half is a straight line corresponding to the liquid surface; see Fig. 4.1. You touch the cell, and make the liquid slosh around a bit, which is seen on the screen by the line becoming wavy until it settles again.

We now start heating up the cell, and by doing that we move up along the coexistence line between liquid and gas, steadily increasing

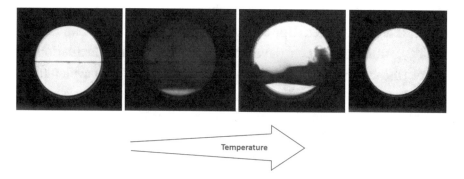

Fig. 4.1 Critical Opalescence: as we increase the vapor pressure and temperature, we eventually come to the critical point after which the distinction between liquid and gas has ceased. At the critical point, the mixture goes completely dark

the vapor pressure, until we approach the critical point beyond which there is no more distinction between gas and liquid. When we reach this point, the substance goes completely black for a moment. Then, as we move through the critical point, the blackness lifts but now the line between the liquid and gas has disappeared. It's neat to see the absence of the line because it shows the phase transition itself. The separation between liquid and gas has vanished above the transition temperature. But what's not only interesting but somewhat startling is the momentary blackness just before the disappearance of the line? What caused this unexpected blackout? No chemical reaction could have occurred—there is nothing to react with. Why was the light blocked from going through the substance? Why were all colors seemingly absorbed by the cell?

A clue is given by what we see just before we reach the critical temperature. On the screen we can see bubble formation—bubbles in the liquid (bottom half), as well as bubbles in the gas (top half). Bubbles in the liquid must correspond to gas bubbles, and sure enough they make their way up to meet the line, which itself is now no longer steady but in constant flux. Bubbles in the gas must be liquid droplets, and indeed they drop down to meet the surface. These bubbles and droplets are precursors to a complete disappearance of any distinction between top and bottom. But as long as the bubbles and droplets are fairly large in size (where we can see them on the projector screen) they do not block the light. Similarly, the atomic fluctuations are too small-scale to much affect light either. But as we approach the critical temperature, bubbles of all sizes starts developing. And when the fluctuations in density (the bubbles) reach a scale comparable to the wavelength of light, then the light is scattered from them. This is actually a general wave phenomenon: roughly speaking, waves are scattered when they encounter obstacles or perturbations on the order of their wavelength. So what the blackout at the critical point, also known as critical opalescence, indicates is that we have density fluctuations at all length scales. It's not just that one color (or wavelength) of light is scattered, it's that they all are.

In the lingo of the field, the correlation length of the fluid/gas goes to infinity at the critical temperature. This is technically not correct, as the sample is not infinite. But compared to the separation of atoms in the liquid, it might as well be considered infinite. Between the atomic scale of about one-tenth of a nanometer and our everyday scale of 1 m lie ten orders of magnitude—ten billion to one. So another way to think of it is to say that fluctuations range from the atomic scale all the way to the size of the sample (here, the cell size).

The amazing thing is that this is also true for other phase transitions, like magnetism. We know that at a temperature higher than what's called the Curie temperature there is no ferromagnetism. So, for example, if you heat a piece of iron to a temperature above 1043 K, or roughly 1418 °F, it ceases to be a magnet. Its spins still have a tendency to align with external magnetic fields impinging upon it, but as soon as the external field is gone, so is any net alignment in any direction. When you cool back down below 1043 K, then a net spin alignment (within a domain) gradually starts to emerge even if no field is present. It happens on its own because spins actually "want" to align themselves with their neighbors. Alignment with neighbors is intrinsically favorable from an energy standpoint.

So why doesn't it always happen, if that's the energetically lower state? A quick answer might be that at high enough temperatures things jostle around quite a bit; the spins have enough thermal energy to easily flip opposite to their neighbors. This hand-wavy explanation kind of makes sense, until you ponder the sharpness of the transition. Why 1043 K exactly? What's special about that temperature? Why should there by a threshold at all instead of a smooth development?

We'll come back to that mystery in a minute. First, let's tie this to our previous example of critical opalescence. At first glance (and second, third and forth glance), the two systems—magnet and fluid—seem very different from one another. In iron, the state that is changing does not pertain to spatial arrangement of atoms; it remains a solid (with the same crystalline structure) through this magnetic transition. Instead, we are dealing with the orientation of its spins, as opposed to their location in space. Nonetheless, at the phase transition, something very similar happens: we get fluctuations—in this case in the spin orientation—on all length scales. And again, we say that the correlation length goes to infinity. If I went in and physically manipulated one particular spin (on one atom in the lattice), perhaps wiggling it back and forth, I cannot say that the effect would be confined to just a certain vicinity around that spin. Even though spins can physically interact with other spins only within close range (spin interactions extend only to nearest neighbors and perhaps to next-nearest neighbors), nevertheless my little manipulation would be felt as far away as you want to consider—as far as a 100 million spin sites away. My little jiggling would be transmitted down the line from spin to spin, all the way out without diminishing. But only at the critical temperature.

At all other temperatures, nature behaves the way we think it ought to. Local fluctuations have local effects and not global ones. As an example, picture yourself driving down interstate I-81 southbound near Harrisburg, Pennsylvania. Presently, you decide to shift into the left lane. One would expect the local traffic pattern to be influenced a bit by your decision, with the car behind you in the left lane adjusting its speed a bit. And perhaps the car behind that car might make a minor adjustment. But we would expect the influence of your decision to fade out beyond that (that is, of course, assuming you didn't cause a major accident and traffic jam). One certainly wouldn't expect a car on I-81 all the way up in Syracuse, NY to be affected by your lane-shift in any measurable way at some future point in time. Things tend to wash out over fairly short distances, typically.

There is one exception to this rule, however, and that is at this critical point between two phases. At this precise point, it is as if my local actions could have effects on all scales. It affects the local traffic, but it also affects people up and down I-81, from Watertown in up-state New York to Knoxville, Tennessee, without diminishing. It's like the system had lost any sense of scale. The car behind me is as close as the car 500 miles further down on the interstate in terms of being affected by me.

If that's not intriguing enough, it gets stranger still. On either side near the critical temperature, the correlation length is already pretty large. As we approach the critical point, either from the left or right, the correlation length *diverges*, meaning it goes off to infinity, as we just said. A similar divergence is also observed in other quantities, but how fast do these quantities rise up? That is captured by a few constants, called *critical exponents* in the literature and conventionally denoted by the Greek letters, α, β, γ, and ν, to name the four most common ones. The larger such a constant is, the quicker the corresponding quantity shoots off to infinity. What's more, many of these critical exponents can be experimentally measured.

Now here's the weird part: for the liquid/gas in SF_6, β is determined to be 0.33; for ordinary water at its critical point (at a very different temperature and pressure) it is 0.33. For many magnets, we get again this number 0.33. Same for other phase transitions, such as order-disorder transitions in alloys. Not only that, but the other critical exponents also mysteriously match up. For instance, the constant ν which described how fast the correlation length diverges was measured at 0.63 for phase transitions in a whole range of disparate systems. Physicists knew this couldn't be a coincidence, and they termed the phenomenon *universality*. Not all systems undergoing a phase transition share the same set of critical exponents, but there are only a small number of "variations on the theme"—a few sets of values also called *universality classes* (Fig. 4.2).

For a while, this is where things stood. The universality of phase transitions had to be treated as an established experimental fact, but why systems so disparate as metal alloys, fluids, ferromagnets, and super-conductors shared common features at their critical points remained puzzling. It seemed especially mystifying given that their respective micro-structures showed no commonalities but actually varied widely. Some attempts at understanding certain aspects of phase transitions turned out to be successful and shed some light on the phenomena, but none of them could explain universality.

Fig. 4.2 The heat capacity of helium-4 near the lambda point, where it goes from a regular fluid to a superfluid. Notice the singularity in the graph at that point. (Image by Adwaele, https://en.wikipedia.org/wiki/File:Heat_capacity_of_4He_01.jpg, licensed under CC BY-SA 3.0)

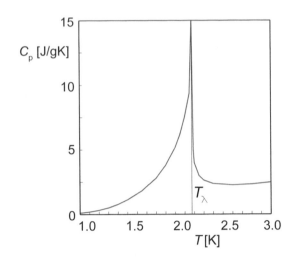

Allow me to briefly describe one such attempt due to its continued relevance in science, but also because it is the simplest way to understand why phase transitions exist in the first place. The name given to this approach is *mean field theory*. As we will see, this theory is quite pertinent in terms of adding to our understanding of emergence in the physical sciences. Even though it's broadly applicable, it's easiest to have a particular physical system in mind. So imagine for a moment a chunk of magnetic material comprised of many atomic spins, arranged on a grid, that can either point up or down, but not sideways. (Such a system is referred to as the Ising-model.) Now, when this chunk is fully magnetized, all the little spins point up or they all point down. When it is partially magnetized, significantly more spins are pointing up than down, or the other way around. Only when the material has an equal balance between up and down spins is it not magnetic. We mentioned earlier that these spins have an incentive to align themselves with their neighbors. Sure enough at very low temperatures, all the spins are lined up and the system has found its lowest energy state. At higher temperatures, some spins have enough thermal energy to flip and anti-align themselves with respect to the rest. At higher temperatures still, no preference between up and down directions can be discerned; things are completely random.

The situation can be likened to the following everyday example. Imagine being part of a jury in a high-profile criminal trial. Evidence has been presented, the prosecutor and the defendant's attorneys have both called and cross-examined witnesses, they have just completed their closing statements, the judge has issued some instructions, and now jury deliberations have begun. At this moment, there is quite some incentive for everyone to come to a consensus and issue a joint verdict. If a clear majority opinion has crystallized, there is strong pressure on you to acquiesce to it. If you persisted as the lone holdout, it would result in a hung jury which would lead to a mistrial, and the entire case would have to be retried. You would probably be blamed by the other jurors and possibly by the judge. You don't want to do this lightly; if there is any way to avoid dissent from the majority, you will find it. If there are a few others siding with you against the majority opinion, the pressure is not quite as heavy but also not entirely absent. And so most juries eventually do find their way to consensus. That's like our spins at low temperature.

Now think of the opposite scenario. You are hanging out with a group of your close friends, as well as some acquaintances and a few strangers mixed in, too, and the topic of discussion turns to local politics. There are different opinions in the room, people have different ideological leanings, different backgrounds or life experiences. At the end of the day, it would still be somewhat nice to leave on an agreeable note, but on the other hand, it wouldn't really harm your relationship with anyone if that didn't happen. A few people might arrive at some common ground, but most others would probably choose to disagree in multiple different ways until the end—no big deal. This is more like the spins at higher temperature.

So how come there is an abrupt onset of agreement as the parameters of the situation are changed from casual social setting to jury trial? Mean field theory has an answer. It postulates that what opinion you have depends probabilistically on the average of the opinions in the room. The *mean* opinion has an affect on you; it makes you more likely to adopt that opinion as your own. How much of an effect it has depends on the temperature (as well as on the coupling strength). But it always has some effect like this. It's not the opinion of anyone in particular that matters. You just care about the average, and so does everyone else. We'll discuss how realistic this rule is in a moment. For now, let's just assume that this is how the game is played. Also assume that we start with a group that has a diversity of opinions in the absence of any social interactions or pressures. So how does this fairly simple model give rise to a phase transition?

Let's examine the two social settings from before again, now in the light of the mean field. In the second scenario of the casual party, there is little social pressure to conform to any majority opinion if there were one in the first place, so a majority opinion will never develop. In the first scenario of the jury, however, the stakes are much higher. If a majority opinion had formed somehow, it would easily sustain itself because individual opinions would be so strongly influenced by it and thus get sucked in. It is easy to see, more or less, how the average opinion would not be zero in this case, but instead strongly aligned in one particular direction. The mere hint of the formation of a group opinion pulls other individuals into it, thus enhancing or amplifying the group opinion.

"But wait, how exactly did the majority opinion form there in the first place?" you may ask. "I thought that the only way a majority opinion could happen is if individual people contributed to it. Now you are saying that the only way people contribute to it is if there is a majority opinion in the first place." There is that feedback, again—a feedback between parts and whole. The parts make up the whole, but the whole influences the parts. What came first? It seems like the classic 'chicken or the egg' problem, a thought loop, an infinite regress from which there is no escape. But there is an escape by way of the following clever procedure.

Let's assume as a starting point (one that we may have to revisit) that a certain strength of majority opinion has taken hold in the group, and to make things concrete, let's say that 70% of the group is of one particular opinion. This number will then influence the opinions of all the individuals separately—the rules of the theory use this number (the 70%) as an input variable to calculate the effect on the individuals (the output variable). So let's say that the theory spits out 65%. In short, given a group support of 70%, you are 65 % likely to hold this opinion yourself. In other words, you are no longer sitting on the fence but are somewhat swayed by the majority.

The catch is that this same rule also applies to all the other people in the group, since everyone is sensitive only to the mean. So now everyone has a probability of 65% to hold this opinion individually. Here is an easy question for you: what percentage of the group would we expect to hold this opinion therefore? 65%! We have come full circle, because we started with an assumption in that measure, but we have arrived at an inconsistency. What we started out with as an assumption is not consistent with the final result. The input does not match the output, 70% ≠ 65%, and we are forced to discard the assumed group opinion of 70% as a viable guess. Instead, we try a new assumption (a little lower) as the group opinion and recalculate the likelihood of an individual holding this opinion based on it. We stop this iterative procedure when we have found what is called *self-consistency*.

It turns out that when feedback between group and individual opinion is weak, in other words when there is little social reinforcement and we don't care too much about what others around us think, then the self-consistent state is no consensus at all, and no majority opinion crystallizes. That makes intuitive sense—why should a cluster form if none of the eclectic individuals have much of an incentive that this be the case? This is like our magnet at high temperature (compared to its Curie temperature). Here this material is unmagnetized with spins pointing in all random directions.

However, when the feedback is strong, as in the case of the jury, then it turns out that self-consistency implies a fairly strong average opinion, a strong majority. This is like our magnet at a low temperature (compared to its Curie temperature). Now this material is strongly magnetized with most spins pointing in the same direction. It stands to reason that there must be a certain feedback strength in between these two examples, where a net majority opinion will first form. This is like a magnet with an ambient temperature equal to the Curie temperature.

In the case of the Ising model of magnetism, the self-consistency condition can be formulated precisely via a mathematical equality that must be satisfied. It's sort of like your typical algebra problem. The unknown x, for which we want to solve, is the net magnetization in this case. The one complicating factor is that unlike what you might encounter in high-school algebra, we are not trying to find the roots of a quadratic equation, say, but roots of a transcendental equation. In other words, we don't just have powers of the unknown x running around in the equation. Instead, what we get is a mix of x and a trigonometric function (the hyperbolic tangent, to be precise) of x. That puts us in a dicy situation, since we are now faced with a mathematical problem that cannot be solved by algebraic manipulation. Mathematics is pretty clear on this—we simply cannot do it by pen and paper. So do we just give up? Not a chance!

Indeed, there is another way—two ways, to be precise. We could resort to numerical recipes to find an approximate solution—a strategy not too different from the educated trial-and-error approach outlined earlier. Or we could try to find the roots graphically. The second approach sounds more fun, but how does it work? Well, we arrange the equation such that on the left side of the equal sign we have the term in x, and on the right we have the hyperbolic tangent of x. Then we plot both sides of the equation separately as curves on a graph, and we look out for intersections. Intersections indicate that both sides attain the same value, and thus they represent solutions to the self-consistency condition. This is shown in Fig. 4.3. The hyperbolic tangent is depicted as the black curve. Also shown (in red and blue) are two straight lines of different slopes, representing the left-hand term proportional to x. Now we notice something interesting. If the slope is greater than 1, the only intersection occurs at the origin, $x = 0$. After that, the two curves just get further and further apart. But if the slope is shallower, then there appears a second intersection.

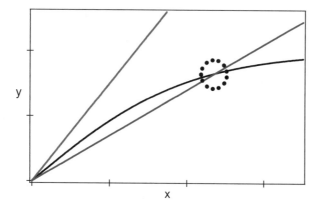

Fig. 4.3 The self-consistency condition in graphical form: it is satisfied at the intersections of the curve (*black*) with a *straight line*. For large slopes of the *straight line* (*red*), the only intersection possible is the one at the origin, x=0. Below a critical slope (*blue*), we get a second intersection—a second solution arises

Now what controls the slope of the straight line? It so turns out that in the model this slope is not arbitrary, but sensitive to the temperature. In particular, the larger the temperature, the greater is the slope. Thus, we can summarize the situation succinctly: if the temperature is too high, the magnetization must be zero (no second intersection). However, when the slope falls below 1 at a characteristic temperature, then another point of intersection presents itself at a nonzero value of *x*. It is this solution that corresponds to the magnetized state.

The bigger lesson here is that the mean-field assumption can explain the existence of a phase transition itself. A new phase of matter arrives due to the changed interplay between the "whole" and its "parts," between a mean property of an ensemble of atoms and the single atom itself. When we have this kind of feedback mechanism across scales, abrupt global changes can occur when the parameters get tweaked. Instead of smooth evolutions of system variables, with feedback we get the possibility of thresholds, the specter of tipping points. In this way, feedback between whole and parts allows for new and separate phases to *emerge*—ones that aren't just continuations or modifications of a previous phase, but which are altogether novel. No trace of them existed in the previous state.

So mean-field theory can teach us about emergence, but before we go too far down that path, let's ask how successful it is as a physical theory? Well, it predicts the presence of divergences (or singularities) in quantities like the magnetic susceptibility and the heat capacity—that's good. But, unfortunately, it often gets the details wrong there. In particular, it is off in the predicted value of critical exponents in many cases. And it gives the same answer no matter what the dimension of the system actually is, whether it be a three dimensional crystal of graphite, or a (quasi) two-dimensional sheet of graphene, or the (quasi) one-dimensional carbon nanotube. The mean-field theory doesn't discriminate between different atomic arrangements, different lattice configurations. In this point, however, the theory is contradicted by experimental data. To nature, it turns out, dimensionality matters a great deal. So what assumption does the mean-field theory make that are unrealistic?

Well, it's probably not that hard to guess. It has to do with the "mean" part. At a party, would I really be influenced only by the mean of all opinions? Or would the opinions of my close friends count more in my mind than the opinion of a total stranger? I think we can all agree that friends usually have more sway over us than perfect strangers. The same is true also for the spins on a lattice. A given spin is much more sensitive to the orientation of its immediate neighbors than to the orientation of any other spin. A far-away spin has no direct influence on it, but it does contribute to the mean orientation just as much as a nearest neighbor would. By averaging over all spins and reducing things to just one number, the mean-field theory smears out the detailed configurations and eliminates the important spatial fluctuations contained in those configurations. A better approach is needed if we want to describe the phenomenon of phase transition more accurately. The physics community was keenly aware of this need, and what followed were a number of refinements to the mean-field approach.

In one spectacular instance, an exact solution was even found for a two-dimensional Ising-model of spins by Lars Onsager at Yale University in 1944. But, while certainly important, none of those refinements, not even Onsager's exact solution of a special case, amounted to the big breakthrough. That would have to wait until 1966, when Leo Kadanoff at the University of Chicago proposed an idea of how to deal with scale and the infinite correlation length of a magnet at criticality. A few years later in 1971, Kenneth Wilson elaborated on this idea, made it a solid calculational tool, and synthesized it into the powerful "renormalization group" formulation, for which he would later earn the Nobel prize. As we shall see, this approach proved highly successful on a number of fronts, including being able to finally explain those mysterious universality classes.

But beyond phase transitions, this approach ended up becoming important for high-energy particle physics as well, as it clarified the physical meaning of a procedure called *renormalization*—initially thought of as just a trick to get around the infinities that were cropping up in quantum field theory calculations. We'll come back to this connection later, as it speaks to the modern concept of the particle. But most crucially for our larger purpose, the renormalization idea will contribute substantially to a more precise understanding of emergence in science.

4.2 Renormalization

Having moved away from the somewhat crude approximation that the behavior of any single spin is influenced only by the mean of all spins, we are left with trying to tackle the more complicated problem: detailed spin configurations and a myriad of local interactions. A spin may only "care" what its immediate neighbors are doing. But now the elegant self-consistency argument from before is not to be had. With the mean-field assumption gone, the relationship between single spin and the group of all spins, the magnet, is no longer clear.

Just as spins in lattices usually have a very limited range of direct interactions, realistic social networks are also highly clustered. We typically listen only to what people in our close network may be saying. Other people will be linked into different social networks. So we still have the possibility of large-scale fluctuations. Sure enough, we see large regional differences in political leanings—the American northeast versus the deep south. But fluctuations in political preference extend also to much shorter scales—sometimes even down to the level of individual neighborhood blocks. This kind of fragmentation on multiple length-scales becomes particularly visible every 4 years with yard signs endorsing either the Republican or the Democratic candidate. It is clear that for most realistic applications, the mean-field approach is too gross of a simplification.

If only there were a way to go from the single spin to the entire group of spins—the macroscopic magnet—in a sequence of smaller steps rather than in one giant leap! Instead of trying to predict my voting pattern based solely on the national opinion polls, let's first examine the political dynamics within my city or county. Instead of considering the entire magnet right away, let's divide it up into smaller blocks containing only a handful of spins. Averaging (i.e., taking a poll) over small blocks has the effect of smoothing out variations that existed within these blocks, and we reduce the block to one number. This also means that they essentially become the new smallest units of our lattice; they become our new "effective spins" [50]. Furthermore, we can now examine interactions at the level of these effective spins, then repeat the same procedure, now with these effective spins, and so on. In this way, we smooth out fluctuations of larger and larger size, but we do it in a sequence of small steps so that we always take stock of what has just happened.

As it turned out, this *spin-block* idea proved to be the key that would fully unlock the mystery of phase transitions. It was that spark of genius which eventually solved the seemingly intractable problem of universality. This story is also a neat illustration of how creativity and intuition are essential factors in theoretical physics. Leo Kadanoff's creativity allowed him in 1966 to look at an old problem from a new perspective. But the idea itself was not going to be enough to see it

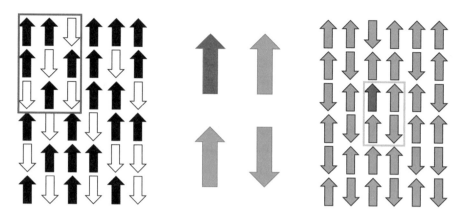

Fig. 4.4 The renormalization procedure in the context of the 2d Ising spin lattice. We average the original lattice (*left panel*) over 3 by 3 blocks, and replace this block by a single spin. This generates a new lattice (*middle*). We rescale this lattice back to the original size and fill in the rest from parts of the original lattice not shown on the left lattice (*right panel*)

through. Also needed was a kind of intuition guided by deep familiarity with the problem and informed by an incisive focus on the important features of the solution. Only then would the initial idea seem worth pursuing, only then could it develop to its full potential.

For a better feel of the creative process, it is instructive to follow Leo Kadanoff's thought process in a bit greater detail. Figure 4.4 illustrates the initial step, where we have taken the original lattice and then grouped sets of nine spins in 3 × 3 blocks. Within each block, we now consider what the majority spin orientation is, and we replace the entire block with a spin of that orientation. That's the middle panel of the figure. As a final step, we shrink the size of the new lattice down so that the spin separations are the same as they were for the original lattice. The end effect is that we arrive at a lattice similar to the original one but with a ninth in number of the original spins. We have just mentally completed one iteration, one mapping, of the *renormalization group*.

So this exercise is all well and good, you may say, but what is the point? What have we gained? It's useful to have the overall goal in mind as we go along, namely we are trying to bridge the gap between parts and whole, between single spin and entire lattice, in a sequence of transformations. The usefulness will reveal itself in due course. Let's take stock of what we have accomplished so far. We again have spins on a lattice, but (a) we have shrunk the lattice size down by a factor of three in both directions, and (b) each spin is really a conglomeration of nine spins. Now comes a highly consequential question: Do these new aggregate spins interact with their neighbors (which are also aggregate spins) in the same way the single spins on the original lattice interacted with their neighbors?

This question can actually be answered rigorously [51]. It can be shown that in general the new spins will interact differently than the old spins did. The coupling constants have to be chosen differently, but also the type of coupling might be different. For instance, if before we had only nearest neighbor interactions between spin pairs, now we might have to include next nearest neighbor interactions in order to make things consistent between the two levels of description. The upshot is that with the redefinition of spin must come a redefinition of the rules governing the spin behavior. In other words, the rules of spin dynamics have to be adjusted as well.

That the rules change should not really surprise us. After all, we just redefined what a spin was. It seems natural to think that this redefinition would have some other ramifications and affect also the interactions (the rules of engagement) between these new constructs. There are two side of the coin: one is that we have essentially reduced the resolution of our image, averaging out small variations and coarse-graining the lattice with each iteration. The other side is that with that coarse-graining must come a transformation of the interaction rules, the spin physics itself. In a sense, zooming out in this manner effectively generates new physical law.

To get away from the abstraction of the previous paragraphs, let's consider the analogy of the party again. Imagine you have a large number of people—it's a sprawling party occupying multiple rooms and spilling out into the porch and backyard—and everyone is bantering with the people around them. Perhaps stretching the analogy a bit, let's say that presently the conversations all turn to local politics. Everyone is talking about whether school taxes should be raised to hire more teachers. Everyone is actively trying to win people nearby over to their view. There are countless arguments being tossed back and forth, and this is going on all across the house where the party is hosted. It is a large dynamical system made up of the individuals at the party and their opinions, if you will.

Applying the renormalization map would be analogous to the following party intervention. The host has had enough of the chaotic chattering, and says that people must form committees of nine members each. Each committee must take a vote and adopt a uniform position on the school-tax questions that had been debated. Each committee must then choose a representative given the authority to argue the committee's positions in front of other committees. Now the interactions happen not at the individual level but at the committee level. We smoothed out variations of sizes smaller than the committee, but in the process we have also changed the nature of the interaction between the new entities we have created—the committees. Committee chairs are not as unguarded in their conversations with other chairs, as they must now frequently consult back with their constituency—the committee members—to make sure the majority is still on board. Of course, we don't have to stop there. If the party is big enough, we can from super-committees, or alliances, and the interaction between these newly created alliances would again be qualitatively different. Indeed, "more is different."

Now recall one characteristic of the critical point of a phase transition. We said that at this point, it is as if the system had lost all sense of scale. Every scale has variations in it, and fluctuations can be coupled across large distances. In that sense the system very much resembles a *fractal*. You have probably seen computer-generated images of fractals—Hollywood discovered a while back that fractals are the way to generate artificial but realistic-looking landscapes. The Mandelbrot set is a prime example, and one of the earliest, of a fractal. On the largest scale you can make out a solid object of a curious shape

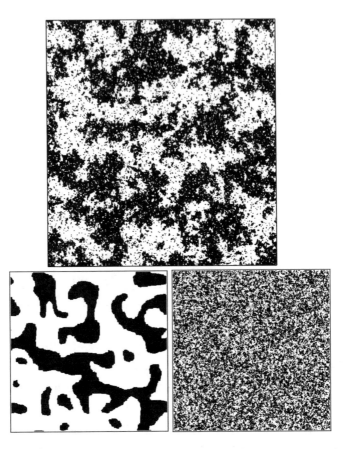

Fig. 4.5 A lattice of coupled spins at various temperatures. I obtained these via a Monte-Carlo simulations outlined in [52]. *Top panel*: The temperature is set to the critical value, and we see hints of self-similarity in the spin patterns. The situation is very different both at lower and higher temperatures, shown in the *bottom panels*

with very fuzzy edges. Then you zoom in to those edges and you see that what appeared to be fuzz is actually comprised of intricate patterns. Within those patterns you see small versions of the original shape. You zoom in on those and sure enough even they have fuzzy boundaries with similar shapes contained in them. You soon realize that the pattern is self-similar on all levels of magnification.

Well, our spin lattice (or our liquid-gas mixture, for that matter) shares the self-similarity feature with this fractal. In Fig. 4.5, top panel, I show the situation at the critical point for a simple spin system.

Spins would intrinsically like to align with neighbors, but there are also thermal fluctuations that "mess up" perfect order. The delicate balance of these two countervailing forces ends up producing a fractal. Here black dots signify the spins pointing up, and white dots depict spins pointing down. We get regions where one spin orientation is dominant, but within these regions we observe patterns of various sizes of the other orientation. For me, the picture communicates visually the essence of a scale-free pattern, where interesting stuff appears on multiple scales. One difference from a true fractal is that with the spins we have a smallest scale—the atomic separation. But apart from that, the spin system does not look qualitatively different when we zoom in or out. Specifically, we can apply the renormalization transformation and coarse-grain, but nothing much happens. It's like we mapped the system onto itself—the before and after shots look identical. Dividing an infinite correlation length by a scale factor is still infinite. It's as if committees mimicked individual people, and alliances mimicked committees.

Notice that this self-similarity only happens at the critical point, however. When we look at a lower or higher temperature, as shown in the lower panels of Fig. 4.5, our eye can immediately pick out the difference. At low temperatures we get uniform domains (black or white regions) forming and they have certain characteristic sizes, and at high temperatures we get almost random patterns, with any order only apparent on very short length scales.

But back to the spin-block idea, where spins within 3 by 3 blocks get averaged together. If we did this at the critical point, and even though we would repeatedly average out local fluctuations while also changing the rules of interaction, it really wouldn't matter. The system approaches a universal pattern—the system's rules approach a generic form. We get a first glimpse that here microscopic details can't really matter much for a system near such a phase transition.

On a basic level, a critical point—by definition—separates two phases of matter and is characterized by that particular temperature at which the two phases meet. Say we wanted to initiate a phase transition.

In the spin example it would be going from unmagnetized iron with fairly random spin orientations to magnetized iron (with large spin domains) as we lower the temperature through the Curie temperature. In SF_6 from before, it would be going through the momentary blackout of critical opalenscence. In a superfluid transition, we would have to pass through the lambda point (see Fig. 4.2). In these examples, we go through the critical point to arrive at the other macroscopic phase. This transit through the critical point is one way in which nature accomplishes the collective reorganization of matter.

But, as we saw, the critical point is also tied to a very generic physical description divorced from microscopic law. We can repeatedly apply the renormalization map without a noticeable change in pattern, and with each iteration we reach physical descriptions that are increasingly generic and non-microscopic. So as we approach the critical point from one side, we can imagine following this process. When we have passed through it and begin to emerge on the other side, we repeat the process in reverse order. There is no reason to think that the rules of microscopic interactions that operated before passing through the critical point would be in any way related to ones in place after the phase transition is completed. Passing through the critical point, in a sense, scrambles the physics—see Fig. 4.6.

This argument can be made much more precise than this somewhat qualitative explanation. But we can appreciate the key insight: the properties of the critical point allow macroscopic phases to establish very distinct physics within them, even though they share the same basic constituent parts. How these parts relate to one another and how they function, however, can be vastly different. Crucially, the physics prevailing in one phase can typically not be derived from the physics governing another phase. The system at the critical point itself (due to its scale-free property) can be mapped via the renormalization procedure to a very generic description which is not sensitive to the microscopic details of phase A. From this state we emerge on the other side in phase B. Since we went through the critical point, we cannot say that phase

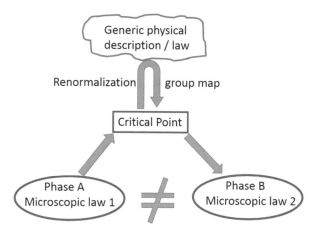

Fig. 4.6 Our system starts out in phase A, but to get to phase B, the system goes through a critical point. The description of the system at this point is not sensitive to the microscopic details of phase A. This means that when we emerge into phase B, its connection to phase A has been severed

A produced or caused phase B. The critical point decoupled the two microscopic descriptions in this sense.

There is one more big result that we obtain via this reasoning: an explanation of *universality*. Why do systems as diverse as liquids, ferromagnets, or superconductors have the same critical exponents, even though on the smallest scale they are really nothing alike. The molecular interactions in the liquid are very different in form from the spin-spin interactions of a ferromagnet, which is different again from the electron behavior in a metal/superconductor. The physics on the microscopic level is all quite different. In the version of renormalization-group formulation outlined just now, what we are doing is combining individual atomic spins into aggregate spin objects, and then repeating the process. At the critical point, this operation changes nothing. We always end up in a place that clearly does not depend on the particular microscopic details of the starting point. We are not even dealing with recognizable physical entities, like spins or atoms, but with generic conglomerations. We always approach the generic level of description, no matter with which microscopic objects or interaction laws we started.

And we approach it always in the same way—hence the common sets of critical exponents. The weird observational coincidence of the critical exponents always attaining certain specific values—something that seemed so mysterious before—is now an actual prediction of the theory.

One of the first things we figure out in life about our physical world is that there are only a few kinds of "stuff." There are these solid objects, which may yield a little but push back when we touch them, and they keep their shape. Then there are fluids, which flow right around our hands, and they take on the shape of the container. Later we may learn that these fluids push back hard when they are compressed in confined spaces, but they yield easily to sheer. From these observations, we might assume that things must be pretty standardized in the microscopic realm. But the opposite is the case. Solids have a myriad of both microscopic structure and microscopic law. There are crystals held together by ionic bonds or by covalent bonds between atoms. There are molecular crystals held together by hydrogen-bonds or dipolar forces. Then there are solids whose cohesion is provided by a delocalized gas of electrons (metals). Furthermore, for any of these microscopic interaction laws, we get a plethora of spatial arrangements and symmetries, such as cubic, orthorhombic or hexagonal. Some solids don't have a regular lattice at all (glasses).

The renormalization idea now explains this basic insensitivity to the microscopic organization or interaction law. After all, microscopic details would have long been washed out in the process of iterative aggregation. Anything that was special or idiosyncratic about any particular starting system will have long been averaged out and become inconsequential. The details of string properties at the Planck scale postulated by string theory have long ceased to matter by the time we even get to the realm of the atomic spin, let alone the realm of spin domains and their organization.

4.3 Phases, Protectorates, and Emergence

An adherent of strict reductionism might say, "In the final analysis, it is always microscopic law that determines macroscopic phenomena." To which we can now reply: "Macroscopic phenomena are often entirely insensitive to the microscopic details? If a whole host of different microscopic systems and physical descriptions can yield the same macro-behavior, in what sense can we say that a particular microscopic law determined or caused the macroscopic behavior? In what sense could we say that the large-scale behavior is reducible to the microscopic law?"

Phase transitions in this way present a strong philosophical challenge for reductionists. It's a challenge that isn't easy to circumvent. And the renormalization group approach is perhaps one of the sharpest weapons wielded by proponents of emergence, for renormalization shields macroscopic phases from the reach of microscopic law. It carves out "protectorates" [5, 53]. These protectorates follow their own rules and have largely decoupled themselves from the details of their substrates.

To illustrate this point, think of a gas. The gas at high temperature is well described as simply a collection of individual atoms; a gas is made up of many, many replicas of the same quantum mechanical object—the atom. So from the gas phase, we can learn a lot about atomic physics, and in fact that's pretty much all an atomic physicist can do. He or she studies gases and their interaction with light, their absorption lines, and so on—almost all we know about atoms comes from gases. But as we cool down the gas, eventually it has to go through a phase transition to become a liquid. And here is where the renormalization idea kicks in. During that process, a transformation takes place that involves self-similarity of scale and in the process de-couples the resulting state from the atomic physics that reigned supreme before in the gas. The physics of fluids is very different—fluid dynamics is not surprisingly a separate branch of physics. The liquid has carved out its own protectorate, its ontological independence from the gas. The self-similarity in turn relied on the correlation length going to infinity. One could argue that it was this singularity embedded in the phase transition that gave rise to the truly emergent nature of the liquid phase [54].

Some of these protectorates have properties that appear to go against their microscopic composition. A case in point is the solid-state phenomenon that goes by the endearing name of the *fractional quantum-Hall effect*. We'll get to that in a bit, but first let's back up a little. In fact, let's back up about 135 years. The ordinary Hall effect is well known to any undergraduate physics student because of its usefulness in measuring magnetic fields easily and pretty accurately. It goes back to Edwin Hall who as part of his physics doctoral thesis at Johns Hopkins University discovered the effect that now bears his name. In 1879 he drove current through thin metal foils in the presence of a magnetic field, and a voltage developed between the opposite edges of the foil that were parallel to the flow of current. In other words, the magnetic field somehow caused a voltage at right angles to the current flow. The more current flowed, the greater was this voltage (at constant magnetic field), and so it seemed convenient to introduce a Hall-resistance as the ratio between these two quantities, the new voltage over current.

The reason that the Hall voltage develops in this context is actually pretty straightforward. The magnetic field deflects the electrons moving through the metal foil, and so they accumulate on one side. This discourages more electrons from joining them on that side, since like charges repel. At this point, an equilibrium is established. The Hall resistance turns out to be inversely proportional to the density of charge carriers in a material, and since this can be temperature dependent (for semiconductors in particular), it thus provides an easy experimental technique to probe electronic densities.

Before we turn to the quantum version of the Hall effect, there is one more significant (for our discussion here) application of this effect. It provides us with a clear experimental proof for the existence of *holes* in semiconductors. A hole is a particle that embodies the absence of an electron, not unlike a bubble in a glass of water. We don't tend to say, "look, there is a spherical region of an absence of water." Or, as the bubble is rising, most of us don't remark that "the water is moving down." No, we intuitively grasp the bubble as a correct conceptual entity in the sense that the dynamics gets easier to describe and predict. In fact, the study of bubbles and bubble instability is quite important in many branches of applied physics and engineering, as a collapsing bubble, for instance, can release an enormous amount of energy and is usually unwelcome (near submarine propellers or in ultrasound imaging, for instance).

In a semiconductors, the analogy of the water are the electrons. Not coincidentally, we speak of a *sea* of electrons. There are an unimaginable number of electrons (on the order of 10^{19} electrons per cubic centimeter) that fill up what is called the valence band of the semiconductor. Since all the states within the valence band are occupied by electrons, as a whole they have no net momentum, no net spin. What would happen if from this electronic background of the valance band, we took away an electron?

Well, it would leave behind a hole that could move through the crystal lattice of the semiconductor, similar to a bubble moving in water. Since the hole (being the absence of an electron) carries a positive elementary charge, a stream of such holes would represent a positive electrical current. And here is where the Hall effect comes in handy. It allows us to distinguish between two situations that otherwise are not easily discriminated: is it negative electrons flowing to the left, or is it positive holes flowing to the right? Both scenarios would give rise to the same overall electrical current and produce the same magnetic field, for instance. But the two scenarios give Hall voltages of opposite signs. Guess what, in the scenario described above we get a Hall voltage consistent with positive charge carriers—holes.

Indeed, we now treat the hole on equal footing with the electron in condensed-matter physics. There are even bound states between electrons and holes in the context of a semiconductor that are reminiscent of the bound state between a proton and an electron, also known as hydrogen. Not surprisingly, these bound states have some properties similar to those of the hydrogen atom.

The quantum version of the Hall effect took much longer to discover—in fact, pretty close to a century longer. In 1980, German physicist Klaus von Klitzing found to his surprise that at low temperature the Hall resistance of his fairly pure semiconductor crystals was quantized. It didn't change smoothly with increasing magnetic field as would be predicted by the classical Hall effect, but showed very discrete steps. When he plotted the inverse of the Hall resistance, namely conductivity, those steps came out to be of exactly equal heights. And the measured height of the quantum steps themselves turned out to be $1/25812.807557\,\Omega$, and thus extremely close to e^2/h—a particular combination of two important constants of nature: the elementary charge e, and Planck's constant h. How close? To about one part in a billion! No wonder the effect was called *exact quantization*. About a year after its experimental discovery, Stanford physicist Robert Laughlin was able to explain the effect theoretically.

Now we are ready to turn to the truly bizarre. At very low temperatures and for extremely pure crystalline samples, we can get the two-dimensional electron gas in these samples to transition into a very delicate quantum-liquid state when very high magnetic fields are present. In this state, the Hall conductivity also exhibits quantum steps as before, but the size of these steps is now a rational fraction of the previous value. The fractional steps were again first seen experimentally, this time by Horst Stroemer and Daniel Tsui. And just like before, Robert Laughlin swooped in to give a theoretical explanation shortly thereafter. All three scientists received the 1998 Nobel prize for this discovery.

The explanation of the effect is what is startling. An electron in the material essentially grabs hold of a single magnetic field line and forms a new particle. But the amazing thing is that these new hybrid particles have fractional charge, such as a third of the electron's charge, or a fifth.

"Fractional charge?", you object. "I thought that all charges came in units of 'e', and this elementary charge is indivisible." And that is still true for elementary particles in isolation (quarks do have fractional charge but can never exist in isolation). Strange as it sounds, the fact is that in this collective system of highly correlated electrons, the new building blocks carry a fraction of the electron's charge. The electrons have morphed into "composite fermions" according to the most recent theoretical description [55], which combine an electron with an integer number of magnetic flux quanta. The description of this quantum phase in terms of particles with fractional charge, such as $e/3$, has since been subjected to rigorous experimental scrutiny. Indeed, there is direct evidence for their existence [56, 57].

But we have to be careful not to misrepresent the state of affairs: it is understood that this direct evidence cannot involve isolating these particles from their environment. It's not like we can take them out of the 2D-layer and examine them in isolation. No, they only exist in the context of the collective phase that has established itself in the physical substrate. The fractionalized charge arises in complete dependence upon the macroscopic phase in which it participates. For this reason such particles are called quasi-particles in solid-state physics; the prefix "quasi" should not be taken to mean that these objects are any less *real*. It simply indicates their relational nature. So, if we can't isolate or separate them, then how can we learn about these fractional quasi-particles and convince ourselves that they are real and not some artificial construct of a theory? That's not an academic question. If we are to take this fractional charge business seriously, there must be some experimental way to get at it.

It is not surprising, therefore, that soon after the theory was proposed, multiple attempts at finding experimental evidence of these fractional charges were launched. One particularly convincing and clever experimental approach has been to look for the typical noise signal associated with current flow, called *shot noise*. This noise comes from the discreteness of the charge carriers. In other words, if one could set up a completely smooth current without any "lumpiness" in charge whatsoever, then this noise would go away. So now we have a way to decide in principle between two possibilities: is the measured electrical current due to N particles of charge e, or due to $3N$ particles of charge $e/3$? The latter should create less shot noise, as it is less lumpy. In such studies (see, for instance, [58]), it could be shown that the noise was consistent with a fractional charge of $e/3$, but not with e.

We need to almost take a step back to absorb the impact. We start out with a gas of electrons confined to two dimensions (a plane) inside a layered material and subjected to a strong magnetic field. This electron gas then coalesces via a phase transition into a quantum fluid which incorporates the magnetic field lines in an intricate way. What we are now saying is that such a quantum fluid is no longer able to be described in terms of the former electrons. Instead, its functional parts are new particles (called quasi-particles) with fractional charge. The phase transition has produced a new collective state of matter, which has eliminated the things that initially entered into the phase transition in favor of new basic constituents. And these new constituents don't even carry the same charge as the original ones!

To be sure, it is not that the electron has divided itself in a kind of 'fission' event. No, we get fractionally charged quasi-particles due to the interplay of a very large number of electrons collaborating and performing correlated motion [59, 60]. It is a radical example that illustrates this larger point: what can even be considered the phenomenological parts depends on the state of the whole. The whole can sometimes define what its own parts are. This example from condensed matter physics is probably as a strong a candidate as one can imagine for downward causation.

4.4 From Crystals to Elementary Particles

So where do all of these discoveries in condensed-matter physics leave us? First, they give us another perspective on the notion of particle. It has become clear that the particle is a very malleable concept. Early on it was recognized that the electron would have somewhat different characteristics in the context of a crystal lattice. Its charge would be shielded a fair amount by the electron sea and the ion lattice, its mass would also be altered by such interactions giving rise to an *effective* mass. In other words, one would never observe the "bare" electron in the crystal, only its renormalized properties. If the whole universe were one giant crystal, then the "bare" electron would be unobservable and therefore, in a sense, non-existent (other than perhaps as an artificial construct of some theoretical models) because it would always be "dressed."

We have seen the acceptance of holes as legitimate quantum particles with measurable effects on the electrical conductivity of semiconductors. These holes are on equal metaphysical footing with these modified electrons. The hole can "orbit" around a negative ion in the lattice to form a bound states very reminiscent of the hydrogen atom. A whole new quantum particle emerges in the form of a phonon—a quantized excitation of lattice vibration. These phonons are essential if we want to understanding measurable thermal properties such as heat capacity and heat conduction of crystals. They can scatter off of impurities in the lattice, and in a superconductor they mediate the binding of two electrons to form a Cooper pair. Again, the phonon is treated as equally real of a particle as the photon by the physical theory; it is explicitly acknowledged that the phonon can only exist in the context of the lattice, but in that context its presence is a phenomenological necessity nonetheless.

Recently, condensed-matter physics has gone even further than that. Instead of showing how, say, electrons get modified when they are inside a solid, we have now seen instances of solid phases where it completely disappears as a recognizable particle. In superconductivity, electrons pair up and morph themselves into bosonic particles with completely different properties irreducible to the individual electrons. Then, these Cooper-pairs *fuse* into a collective quantum state. A similar thing happens in helium-3, as we have seen. In semiconductor hetero-structures, we get two-dimensional electron gases that in the presence of a strong magnetic field condense into a quantum fluid described by fractionally charged quasi-particles. In high-T_c cuprate superconductors, which are still not fully understood even in the face of increasing experimental characterization of them, there have been suggestions by theorists that what is happening with the electrons is *charge-spin separation.*

In these cuprate systems, excitations appear that carry only charge but no spin, and other excitations that carry only spin but not charge. This is odd, as the electron is a particle with charge and spin combined into one package. How could these two aspects of an electron separate? It's sort of like the cup in front of you dividing up into its shape and its color, with the two moving independently. This provides yet another way in which the bare electron can be *fractionalized* or otherwise deconstructed inside matter. Another example in this vein would be the appearance of magnetic monopoles inside certain honeycomb ferromagnets, also known as *frustrated* magnets. Magnetic monopoles, you may know, are generally forbidden in electromagnetic theory, but in the context of these special magnets, they seem to arise in the form of quasi-particles that can move freely about the honeycomb lattice.

So what is it with macroscopic states of matter that makes particles behave so strangely? Why are things getting so weird in the realm of crystals and fluids? A case can be made that the situation in contemporary condensed-matter physics is not so radically different from that of elementary-particle physics. Elementary particles are not studied within the context of macroscopic matter, instead they are studied by themselves in isolation. To truly isolate them from anything else, we have to put them in a vacuum. Then it is certain that nothing will disturb them, and we can study them as they truly are within themselves, right?

Wrong. The thing we need to remember is that the vacuum is not nothingness according to modern physics. To the contrary, it is teeming with virtual particles and anti-particles which can come into existence momentarily and then annihilate soon thereafter. To put it jokingly, the vacuum isn't at all what it used to be. In fact, the Heisenberg uncertainty principle specifically allows this kind of popping in and out of existence. At very short time (or length) scales, the uncertainty in energy (or momentum) becomes very large, so large indeed that we can accommodate the creation of mass-energy. The upshot is that "bare" elementary particles are also unobservable, since they get modified, or renormalized, by the presence of the vacuum's virtual particles. These virtual particles then respond to the presence of the original particle of interest.

Take for instance the electron. In the early days of quantum electro-dynamics, or QED for short, scientists were confronted with infinities in their calculations of basic quantities. Infinities are never good in a calculation. When expansion series do not converge, it is usually an indication that something has gone wrong, that some assumption or approximation isn't valid. QED was supposed to be the final quantum theory of light and matter, so these infinities were somewhat of an embarrassment. The formal solution was the renormalization idea in the 1950s by physicists Murray Gell-Mann and Francis Low.

The upshot was that yes, the bare electron's charge might be infinite, but if we probe the electron charge at a reasonable distance, then we get a finite value due to the response of the vacuum. In other words the electron has a screening cloud around it at small length scales. If we probe the electron at scales much beyond the so-called Compton wavelength for the electron, its charge will be found to be e. But if we were to pierce the shielding cloud and probe the electron's charge at a closer distance, the measured value would go up. The infinities that the theory predicts are not so worrisome because the bare electron cannot be observed anyway—the vacuum will always shield it, and so above distances comparable to the Compton wavelength, we get something finite.

Nevertheless, in the formal renormalization procedure, an arbitrary length scale had to be introduced at which we were first going to probe the electron. Ken Wilson's renormalization group approach finally removed the seeming arbitrariness of the length scale that was inputted into the previous procedure. Broadly speaking, the interpretation now is that the detailed physics below a certain length scale doesn't matter for a theory that aims to describe the world at a particular length scale. The microscopic details get washed out and we get an effective theory at the length scale we are probing.

So, coming back to QED, at very small scales new physics must be present, but it is not essential to know this new physics above a certain scale where QED takes over. Why? Because QED is insensitive to those details. Many different versions of the new physics beyond QED at smaller length scales would all converge to QED. And that new physics might yet again only remain valid down to another, even smaller, scale—below it, another new theory would take over. But, according to the renormalization idea first introduced to understand phase transitions and universality, knowing the new theories at ever smaller scales is not important, as each scale's physics is insensitive to the details of the smaller scale. At each scale we have effective theories that become their own protectorates. This is a very profound view of emergence that contemporary physics has uncovered. In the words of Robert Laughlin,

> What we are seeing is the transformation of worldview in which the objective of understanding nature by breaking it down into ever smaller parts is supplanted by the objective of understanding how nature organizes itself. [...] The fractional quantum Hall effect shows that ostensibly indivisible quanta - in this case the electron charge e - can be broken into pieces through self-organization of phases. The fundamental things, in other words, are not necessarily fundamental. [5, p. 76]

Chapter 5
Beyond the Linear Approximation

5.1 Chaos Theory

If you happened to be good at math in high school, chances are you liked the subject. If you're naturally gifted with a racket in your hand, it wouldn't surprise me if you enjoyed playing tennis or squash. Personal success, it seems, focuses the mind like little else. Sometimes this rule also applies to entire disciplines. A perusal of old physics textbooks, for instance, quickly suggests that our discipline had been harboring a love affair with all things *linear*. Not that linear systems are very common in nature. No, linearity is the rare exception and not the rule. Rather, fact is that physics has had a lot of success with linear systems. They yield to mathematical analysis, and this analysis in turn allows for precise predictability. It's not surprising then that in physics linear systems soon became the model children, the teacher's pets. In textbooks they received most, if not all, the attention.

Take the example of a mass hanging off a spring—a system drilled into every physics student. When disturbed, the mass performs up-and-down oscillations, and the frequency of this oscillation can be mathematically derived. Why? Because we assume that the spring is linear. Stretch the spring by a certain distance and it pushes back with a certain force. Stretch it twice as much, and the force is also doubled. We

© Springer International Publishing AG 2017
L.Q. English, *There Is No Theory of Everything*,
DOI 10.1007/978-3-319-59150-6_5

say that the force is proportional to the stretch (within limits, of course). Material science also often assumes this kind of proportionality. The strain experienced by an elastic material is proportional to the stress exerted, and quite a few material properties can be derived under this assumption.

Linear systems exhibit a delicate property: their response is exactly proportional to the external stimulus. Double one and you double the other; triple one and (surprise) you triple the other. Subject the system to two different types of stimulus simultaneously, and its response is a simple superposition of the single-stimulus responses. This ability to decompose stimuli and reconstruct responses is what makes linear systems so nice to analyze. The superposition principle obeyed by linear systems yields one other really nice feature.

That is, noise is manageable. Noise is, of course, a fact of life. Radio waves had to travel miles through the atmosphere before they were picked up by your antenna. Not surprisingly, they accumulated a certain level of noise which embedded itself in them. The linearity in your RLC resonator and amplifier guarantees that your speakers will produce nothing more than a little background static—that's all. Static can further be reduced after reception using filtering and signal post-processing, so most of the time it represents only a slight inconvenience. The station still comes through loud and clear, and the voices on your favorite NPR program are not garbled.

In systems that are not linear, noise is often much less benign and much more consequential. The realization that small fluctuation—noise—can sometimes radically alter the evolution of nonlinear systems had to await the 1960s and 1970s, when it finally dawned on the scientific community that, in their fixation with linear systems, they had missed key aspects of natural law. Much of the early credit goes to Edward Lorenz—an MIT meteorologist and mathematician interested in bringing a quantitative approach to weather forecasting. Lorenz was exploring simplified models of weather numerically with the help of a rudimentary computer. At one point, he wanted to reproduce a previous data run, but instead of entering the full six-digit number as his initialization, he rounded to three digits [18]. The difference should have been small, but Lorenz soon discovered that after pretty decent initial agreement, the new data began to diverge rapidly from the old one. As the story goes, Lorenz had a Eureka moment, deciding on the spot that long-range weather forecasting must be doomed [18]. And so the "butterfly effect" had been born—also known (in more technical terms) as 'extreme sensitivity to initial conditions.'

Even fairly simple systems exist that starkly exhibit this extreme sensitivity. The double pendulum is one of my favorite examples. Here is a basic mechanical contraption—a pendulum, usually in the form of a metal bar, at the end of which is attached another such pendulum. Let this double pendulum go from a sufficient height, and the resulting dynamics becomes rather erratic. The bottom pendulum, in particular, performs some unexpectedly wild motion: sometimes it is being flung around its pivot (at the end of the upper pendulum) in rapid succession, and at other times it abruptly changes course and spins in the opposite direction. Even towards the end, when friction has dissipated much of the energy, the motion still has one last surprise in store for the observer, managing to fling the bottom pendulum around when you didn't expect it anymore. It is a mesmerizing spectacle to watch. Energy shifts fluidly from one pendulum to the other, with each pendulum acting as the stimulus for the other, each responding to the other's motion, and both passing fluctuations—large and small—back and forth between them (Fig. 5.1).

The motion is as surprising as it is intractable. It is intuitively obvious that there is no way of understanding the motion from the perspective of two individual pendula. In fact, the single pendulum never displays anything beyond boring periodicity—a result that can be proven mathematically. But couple two of them together, and all bets are off. From a dynamical perspective, then, the whole is more than the sum of its parts. The whole exhibits types of motion that are completely unattainable by the parts in isolation. Only when two pendula are coupled together in this manner, can chaos arise. Only this coupling allows each pendulum to exhibit chaotic dynamics, to

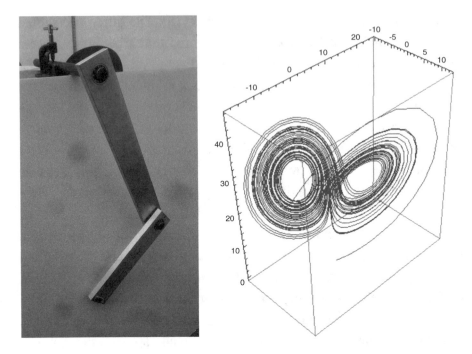

Fig. 5.1 *Left*: The double pendulum exhibits chaotic behavior when released from a sufficient height. *Right*: The strange attractor associated with the Lorenz system. The trace never crosses itself and never repeats, but stays confined nonetheless

move truly aperiodically without ever repeating. Only then can each pendulum move along what is called a *strange attractor*—a geometric object, if you will, of finite extent but of infinitesimally fine structure.

In spatially extended systems, such as a fluid running down a pipe or the wind currents making up a weather front, it is as if an infinite number of such pendula had been coupled together. Each parcel of air might be represented as its own little pendulum, passing energy back and forth with neighboring parcels. Thus, fluctuations that happen to manifest in one small region of the system—a parcel of air, say—can be passed seamlessly to other regions, causing disturbances there. Disturbances can combine with other fluctuations to produce larger-scale turbulences. Every part of the system is sensitive to what happens in every other part. Small motion initially confined to some isolated corner can ripple out and start influencing the subsequent dynamics of every other part of the system.

A strategy of mentally breaking the system apart into smaller units, and then examining one unit at a time, is destined to fail. What goes on in one part of the system does not die out, it reaches other parts, and due to the afore-mentioned extreme sensitivity flings them into new trajectories. In this sense, there are never any dull moments in the life of a complex system. Every moment in time at every point in space is a critical point. The situation can only truly be approached holistically.

We intuitively grasp our human condition in these same terms. Every moment in time is of existential criticality. As individuals, we are connected to so many other people, whether we happen to know them or not. In traffic, we are affected by the split decisions of hundreds of fellow drivers and pedestrians. At a job interview, even a small hand gesture or facial expression on our part could make the difference between us getting the job or being passed over. A chance encounter with a stranger could make the difference between us accepting that job offer over another. Then, of course, these decisions deeply affect the subsequent trajectory of our life—most immediately, it influences where we will move and who we will meet next. And so does every other decision—even minor ones. Indeed, every action, however small, ripples out and affects the lives of all those around us. If we move closer to our newly accepted job, perhaps our kids will have to switch schools, which in turn enables a myriad of new interactions and terminates a myriad of others.

On another level, as individuals we are also highly sensitive to the broader currents constantly emerging within social groups, as well as to large-scale societal patterns. People conscripted to fight in a war serve as a stark reminder of large-scale political circumstances and events acting back on the lives of individuals. In all of these examples, it is impossible to treat external interactions as small perturbations that do not substantially affect the evolution of individual parts. As we have already seen before, strongly interacting systems generate new sets of phenomena that are fundamentally irreducible.

In previous chapters, the "more-is-different" thesis was supported and developed in ways that shared many of the same overall features, but where each new physical context also added its own unique perspective. Chaos theory now emphasis yet another dimension of emergence—one that raises the question of strict causality. A popular rendering of the *butterfly effect* has it that the flap of a butterfly's wing in Australia can cause a hurricane down the line in Florida. It is true that a system's dynamics is globally sensitive to even the smallest of perturbations, and this property of chaotic dynamics then somehow naturally conjures up butterflies. But the flip-side of chaos is that complex, strongly-interacting systems and networks do not allow us to sort out cause-and-effect relationships. They make it impossible even in principle to say that event A caused event B to occur, precisely because we cannot predict these systems and evolve them forward in time. All we can say is that if event A hadn't occurred, then the detailed evolution of the system would have proceeded differently.

Can we be certain that event B would not have happened in that scenario? No! The fact is that the tornado off the coast of Florida is a collective phenomenon of the atmospheric system. It cannot be traced back to the behavior of a few isolated parts of the system. Yes, there are mathematically precise rules operating locally everywhere within the atmosphere, but these rules do not allow us to predict how the system as a whole behaves on a large scale. We simply cannot ferret out a list of local events that formed the basis of a subsequent phenomenon. To put it loosely, everything was somehow involved in bringing every subsequent thing into existence.

I am reminded of the ancient Greek dictum that you cannot step into the same river twice. We simply have no way to run the experiment of life a second time, this time stopping that butterfly from flapping its wing at that precise moment. If we could stop the butterfly (and affect absolutely nothing else at all), then wait a month to find the hurricane not materializing, in that case we would have established a cause-and-effect relationship empirically. But we cannot do that. And so, without either any empirical method or a theoretical prediction at our disposal that could establish such a relationship, we have really no choice but to surrender. We are forced to abandon the notion of strict cause and effect at the level of these large-scale macroscopic phenomena. The equations of physics operate underneath it all, but in chaotic systems they give rise to global evolution that defies the clear discrimination of causality.

In light of emergence, this also means that the causes of a phenomenon within a spatially-extended and strongly-interacting system cannot be traced back to the many local occurrences in isolation. The connection between what happens locally and what will happen globally is irrevocably lost. As the complexity of systems increases, strict causality becomes increasingly difficult to establish within them. As we will now see, this state of affairs also extends to nonlinear systems that, instead of exhibiting chaos, give rise to the formation of spontaneous patterns—a new kind of nonlinear order.

5.2   Self-Organized Patterns

Anyone who has ever seen a hot air balloon lift off the ground and float into the sky understands on a visceral level that hot air rises. When the air in a hot-air balloon is heated up by the flame underneath, the air expands and becomes less dense; buoyancy then drives the air upwards, and with effortless grace it carries along with it the balloon and the people in the basket attached to it.

The basic phenomenon of warm air rising happens, of course, also naturally. When air at low altitudes is heated sufficiently by the sun (or by a surface that is hot due to the sun), there can be upstreams or thermals as a result. Paragliders often take advantage of these thermals to soar high into the air. The colorful canopies circling around an invisible center is quite a memorable sight. In fact, it takes particular skill and practice for a paraglider to home in on and ride along these thermals, since whenever there is warm air rising, there usually is cold air sinking not too far away. That's because the rising warm currents displace colder air which then has to go somewhere. The technical name for this phenomenon is *convection*. Not surprisingly, it plays an important role in meteorology—the rising air is often quite moist, but the water vapor condenses out upon cooling, leading to rain fall.

The basic phenomenon of convection had been known for some time, but it was French physicist Henri Bénard around the turn of the last century who first studied it carefully in a laboratory setting under controlled conditions. As is often the case in science, meticulous experimentation and careful inspection of a known effect can reveal unexpected gems.

Henri Bénard had stumbled on the phenomena that now bears his name by random chance [61]. As a Ph.D student at the Collège de France in Paris, he was handling molten wax (a dielectric) into which some graphite dust had been mixed for applications in radio-wave detection. In the wax he noticed regular patterns, which piqued his curiosity. What caused them? Nobody seemed to know. Had this phenomenon ever been studied before? The answer seemed to be no. And so began Bénard's dissertation project. It was a simpler time in science.

What Henri Bénard now did over the next couple of years was essentially to confine various fluids within shallow containers, carefully heating them from below. He went to great lengths to make sure that the bottom of the fluid was heated as uniformly as possible—he wanted to avoid artifacts of non-homogeneity. He was extremely careful in other ways too. For instance, he used optics techniques to determine fluid heights and motion extremely accurately. For one particular liquid layer with a thickness of only 1 mm, he was able to detect variations in the top surface at a resolution of 1 μm, roughly a tenth the width of a human hair [61].

It was a work of remarkable experimental skill and exacting precision. Yet, in the end, his Ph.D. committee was not entirely pleased. Considering how important Bénard's discoveries would later become, it is more than a little ironic to read the verdict of the professors: "Though Bénard's main thesis was very peculiar, it does not bring significant elements to our knowledge."

What was his thesis? We can begin by picturing Bénard's setup. The bottom layer of the fluid is made hotter than the top, so there is now a vertical temperature gradient across the fluid. Where else do we encounter temperature gradients in our daily life? Across the outside walls of our house in winter, for example! And no matter how much insulation we have in those walls, the response to this temperature gradient will be some amount of heat conduction through the wall from the hot to the cold side (although insulation does, of course, reduce this heat flow in magnitude).

So this is one thing that could happen. The fluid could remain at rest, just as the wall of our house remains static (hopefully!), and heat could simply move through it from hot at the bottom to cold at the top. Indeed, this is sometimes observed. But, interestingly, it only occurs below a certain threshold in temperature difference! As the temperature gradient gets larger by upping the temperature at the bottom, something else happens entirely. Convection starts setting in—meaning that the fluid itself will start moving. Hot fluid at the bottom will rise, and cool fluid at the top will sink.

The net effect is again to transport heat from bottom to top, but this time not because of heat conduction, but because the fluid itself moves. One can easily imagine that this method would result in a lot more heat transport; it is simply more effective in shuttling heat energy from hot to cold. Rather than the heat having to diffuse through the material by molecular collisions, the material moves on a macroscopic scale. Bénard's great discovery was that for a variety of different fluids there always was this sharp onset of convection—below a critical temperature difference heat transport was dominated by conduction, but above it convection took over.

A practical example is the double-pane windows that are so much better insulators than single pane windows (even of twice the thickness). Why do they work so well? The answer is that static gases, such as air, are very poor conductors of heat. This is particularly true of the noble gases Argon or Xenon that are now commonly used in windows. More specifically, if these are trapped between the panes of windows, they provide good thermal insulation. So your next thought might be: if a gas is such a good insulator, then why not increase its thickness; why not separate the two panes by a wider gap and fill the larger volume with more gas. Shouldn't that be even better? After all, that's essentially what people do with fiberglass insulation in the attic—the thicker the layer, the better the insulation.

This rule fails, however, for windows! The thickness of the gas layer cannot exceed a fairly small value on the order of one inch. Beyond that, convection would commence in the gas between the panes, and the insulating properties of the window would be destroyed. In winter, the hot gas near the inside of the window would physically find a way to travel to the outside, forcing cold air near the outside to move to the inside. It would almost be as if there was no barrier at all.

Let's return to Bénard's original setup—a layer of fluid heated from below. As with most problems in science, we can always dig deeper, and indeed our discussion has so far glossed over a few important details. When you think about it a little, the entire bottom layer of fluid would "want" to move up. But then a second thought enters: the entire layer can't rise; it can't physically happen like that. The top layer of the fluid must have a path downward to replace the hotter fluid rising. Otherwise, the entire fluid would start levitating leaving underneath it a vacuum! So it quickly becomes clear that the bottom layer cannot move up in complete unison. The fluid flow has to organize itself spatially into upward channels and downward channels. If convection is to proceed, a *breaking of the symmetry* is inevitable. If traffic is to flow on a road, an analogous symmetry breaking into discrete lanes has to occur.

The symmetry is embodied in the statement "the entire bottom layer of the fluid wants to rise." Apriori, there is nothing that would favor one region of that warm fluid layer over another (assuming the container is large and any vertical walls are far away). Yet it is also clear that this symmetry has to be broken if convection is to occur. There will have to form spatial distributions or patterns of rising and falling currents. Some parts of the container will have been selected as carrying warm fluid up, and other parts will have been selected as carrying cold fluid

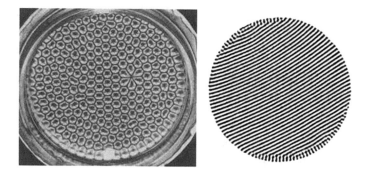

Fig. 5.2 Convection as seen from the top. *Black* (*white*) shading represents regions of falling (rising) fluid. On the *left*, a hexagonal organization of up- and down-moving fluid. From [62]; also reprinted in [63, p. 863]. *Right*, the more ubiquitous arrangement of convection rolls. Image from [64]

down (Fig. 5.2). On what basis does this selection occur and by what mechanism? And what spatial patterns are actually observed?

Let's tackle this last question first. When Bénard first conducted these experiments, he saw beautifully regular patterns of rising and falling fluid currents. In fact, the system seemed to prefer a hexagonal periodic arrangement not unlike the honeycomb geometry of bees. Later, it became clear that the quintessential pattern that forms near the critical temperature difference are rolls of a certain width. Viewed from the top, these periodic rolls look like a plane-wave of a characteristic wavelength—stripes of dark and bright. In round containers, those stripes can sometimes bend to form rings or even spirals around the center.

What's more, Bénard found the rolls and their associated wavelength to be independent of the type of fluid used. In fact, modern experiments don't use fluids at all, but rather compressed gases (often near their critical points) and the phenomenon persists. We always see rolls whose diameters are nearly equal to the height of the convection cell (defined by the separation of the cold and hot plates). The question is what ties all of these detailed observations about convection together, and what explains the regularity across different media?

A few years after Bénard published his findings, it came to the attention of the eminent British physicist Lord Rayleigh, and he immediately set out to find a theoretical explanation of it. This was the same guy who had already explained why the sky was blue (Rayleigh scattering of light from molecules). So in his usual manner, it didn't take him long before he succeeded brilliantly. In fact, the theory was so crisp and insightful that the phenomenon is now referred to as Rayleigh-Bénard convection. But what was Rayleigh's starting point? How is it possible for physics to explain the abrupt onset of convection as the temperature difference is gradually increased? How can one predict the size of the convection cells from basic physical law?

The theory of convection, it turns out, does not start with microscopic law. It is firmly based in the macroscopic laws of fluid dynamics. Quantum mechanics does not enter; neither does atomic or molecular physics. The modern starting point are the so-called Navier-Stokes equations for fluid flow, which deal with macroscopic measurables such as pressure, density, volume, temperature, and other thermodynamic quantities [63]. It is very clear that the description of the fluid is a thoroughly macroscopic one. There is no reference to atoms at all; in fact, the fluid is assumed (for all intents and purposes) to be a continuous medium.

Indeed, microscopic details are conspicuously absent. Rayleigh showed that the only thing that mattered for the onset of convection was one number, not surprisingly called the *Rayleigh number*. Nothing else about the fluid mattered. Only a particular combination of macroscopic fluid properties, such as its thermal expansion coefficient, its thermal conductivity and viscosity, entered into the picture. One type of fluid would behave exactly like any other as long as this combination of macro-constants was the same. The theory didn't even care about what atoms or molecules the fluid was actually made up of. Yet, it correctly predicted the critical temperature difference, the type of pattern that would emerge past it (rolls, see also Fig. 5.3), and the width of these rolls. The roll width, it turns out, is very nearly equal to the thickness of the fluid layer, which may be on the order of an inch or so (a few centimeters). So that's the characteristic length-scale of the emergent

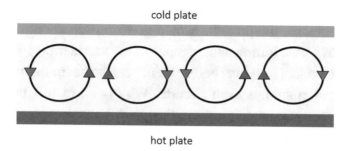

cold plate

hot plate

Fig. 5.3 Cross-sectional view of convection rolls—hot fluid at the bottom rises, cools, and then sinks. There it heats up again, and the process continues

pattern. The characteristic length scale of a molecule, or the inter-molecular separation, in the fluid is on the order of a few tenths of a nanometer—about 100 million times smaller. The two length-scales are vastly different.

So allow me a brief interlude, if you will, where you imagine that you are a water molecule caught up in one of the up-currents associated with a convection roll. You notice that the pressure is creeping down, and you ask your neighboring molecules if they know what's happening. Nobody around you seems to have a clear answer, but they all confirm that they have also noticed a slowly declining pressure. Reports are coming in from the farthest reaches of the known fluid, and they tell of the same thing. Everyone's wondering why this large-scale change may be happening, debating a cacophony of explanations, until a particularly precocious molecule, going by the name of Albert, announces a radical theory. "We are all moving up! That's why the pressure we've been measuring is gradually getting less and less." He explains that according to his theory, the pressure was the result of gravitation and the number of other molecules above them. A lessening pressure meant fewer molecules above them, which in turn implied that they were all moving closer to the edge and would eventually reach the top.

There is a stunned silence for a moment, then a flurry of questions: "What do you mean by the 'top'?" and "How long before we get there?" Albert says that the 'top' would mean the edge of their 'universe'—a place difficult to imagine. And he declares that according to his calculations, the pressure would go to zero in about two billion nanoseconds. "Two billion nanoseconds? That's an eternity! Some of us will long be broken into ions before then!" can be heard multiple times. A final question stumps even Albert: "Why is all of this happening? What can we expect when we get to the top?" It just doesn't make sense to any of the molecules; no matter how hard they try, nobody can come up with a plausible explanation for why they all seem to be moving in the same direction towards the top. And no-one dares to venture a guess as to what will occur afterward.

Albert, however, has been wondering. Where did all those molecules go that used to be on top of them? Could they have all left the group, evaporated as the term went? That was certainly a plausible explanation, but according to his earlier calculation it couldn't account for the speed at which the pressure was changing. Something else must be responsible. The only alternative seemed to be that these other molecules at the top were sinking. But no reports far and wide had recorded sinking molecules. And now gradually it dawned on him. After a few more calculations, he was ready to make his second ground-breaking announcement.

"Our universe is not static. Instead, we are caught up in a cosmic vortex. The diameter of this vortex is roughly two hundred million nanometers. If my theory is correct, we will reach the top, then briefly go sideways, before embarking on a long descent."

You can probably guess the moral of this silly fable. The directed motion of single water molecules are determined by macroscopic laws and processes. The motion can't be understood microscopically (except by our precocious molecule). It's an instability in the fluid as a whole that gives rise to the formation of coherent structures on a global scale. It is these collective patterns, the convection rolls, in which the microscopic constituents of the fluid (the molecules) are caught up and participate. The molecular motion here is partly dictated, and certainly constrained, by processes that can only be comprehended at a higher level. The pattern selection and formation represents a strong form of *downward control* on the molecular level.

The axial shape of these rolls—whether they are straight or round—and the number of rolls is also determined by boundary conditions, like the container shape and size. Again, the pattern is sensitive to macroscopic details (such as the container) and not microscopic ones. Again we see that the individual molecule's "behavior is contextually constrained by the global structure which it constitutes and into which it is caught up" [65].

No one is claiming that the larger structures force molecules to do something that is in violation of microscopic laws. Quantum mechanics is not violated by the collective motion of molecules, but it also doesn't contain in it the reason for that motion. That reason is found at a higher level. It is a statement about self-organized patterns and downward control that contains yet another novel facet of emergence as it is revealed in the physical sciences; it paints a clearer picture of what is meant by emergence. Neurobiologist William Newsome's gets to the heart of the matter in the following definition:

> By "emergence" I mean that complex assemblies of simpler components can generate behaviors that are not predictable from knowledge of the components alone and are governed by logic and rules that are independent of (although constrained by) those that govern the components. Furthermore, the intrinsic logic that emerges at higher levels of the system exerts "downward control" over low-level components. [66]

Indeed, the "logic and rules" governing societies are, for all intents and purposes, independent of those governing neurobiology. There is no direct connection between the two systems of rules—one cannot transition smoothly from one to the other and back. Of course, societies are comprised of people and people's thinking is described by rules of psychology. Psychology presupposes a human brain as the substrate. So, of course, societies wouldn't exist without people coming together, and people wouldn't exist without cells cooperating, with some of these cells specializing to become neurons. Naturally, higher-level rules depend on higher-level systems comprised of parts, and these parts operate according to their own rules. So, an indirect connection exists by

Fig. 5.4 Examples of simple regular patterns in nature. Their origins often seems somewhat mysterious. (Clouds by Unasia9, https://commons.wikimedia.org/wiki/File %3AAltocumulus_radiatus_clouds.jpg, licensed under CC BY 4.0. Sand picture by: Rosino on Flickr, https://commons.wikimedia.org/wiki/File:Morocco_Africa_Flickr_ Rosino_December_2005_84527213.jpg, licensed under CC BY 2.0. Zebra picture by Derek Keats, https://www.flickr.com/photos/dkeats/16906289242/, licensed under CC BY 2.0)

virtue of mere composition. But no direct connection between systems of rules can be formulated. Neurobiology and sociology are separate fields of study for a reason.

It may strike you as obvious, but it is important to recognize that pattern formation really is all around us (Fig. 5.4). Think of the simple striped patterns that spontaneously forms in clouds and sand, for instance, or the striped or dotted markings on some animals' skin. In all these instances, it is an instability in the steady, uniform state that amplifies particular fluctuations from which recognizable patterns then arise. So convection is not some esoteric trick of nature; it is a neat illustration of pattern formation more generally. To explain the beautiful skin patterns on animals from zebras to anglefish, *reaction-diffusion* models are the typical starting point for analysis [67]. The term reaction-diffusion here refers to molecular processes involved in the placement of skin pigmentation. These molecules react with local agents, but they can also spatially diffuse into neighboring regions of lower concentration. Once the details of such a mathematical model are set to reasonably approximate the biochemistry involved, the general mode of analysis proceeds as before.

Come to think of it, we don't have to stop at simple realizations of geometric patterns in nature. What about complex organizations we see in the biological world? Biology, of course, is the study of highly structured organization, from single-cell organisms to redwood trees. If "downward control" is responsible for the simple patterns of convection rolls and the like, it stands to reason that it may play an even larger role in biology.

5.3   Displays of Temporal Order

One of the nice things about being at a smaller college is that sometimes you can just have fun with research. The pressure to publish and to find streams of funding is not nearly as intense as it is at a large research university. Things don't always have to lead to major publications or be funded by a grant agency. Sometimes it's alright to explore a fun idea you've come up with on the fly. So, a couple of years ago, when two physics majors—Ashley and Jean-Marie—asked me if I had any fun projects for them to work on, I said: "Sure!"

It turns out that I had just finished reading a book on social networks, and I had this idea of building our own mini-version of a social network that was simple enough that we could know everything that was happening. So after throwing some ideas around, the three of us settled on the final plan: we were going to build an electronic network with eight people in it. They would communicate via pushing buttons and activating LED lights at the other people's stations. So whenever someone would push their button, lights would go on at everyone else's stations. We would give the participants the stated goal of pushing their buttons at a rate that felt natural to them, but also to try their best to come into synchrony with the lights they saw flashing before them from the other participants' buttons. So the question was, could this group of eight participants globally synchronize their responses?

Perhaps a month and a half into the project, Ashley and Jean-Marie had built a first version of the circuitry. As you can tell from Fig. 5.5,

Fig. 5.5 The (slightly messy) wiring network designed to test synchronization in the collective responses of participants

it was sort of a wire mess, but it worked. When we ran some actual tests with volunteers we could round up in the physics majors' room, it was quickly clear to us that the answer was yes, they can. Within less than 20 s, groups usually managed to sync up. If you have ever heard a concert audience ask for an encore by clapping rhythmically in unison, this outcome probably doesn't surprise you much. We know that people can organize their clapping in this manner even without a conductor directing it, and so one would assume that people can do the same in response to a visual stimulus. But the question we really wanted to answer was the following: how long would this phenomenon survive as we turned off connections in this network.

To answer this more interesting question, we had designed the communication network to be able to go in, flip one of the DIP switches, and disengage connections one by one. So, that's what we now did; we took out connections starting from the fully-connected network of everyone seeing everyone else's response. What was interesting was that global synchronization disappeared even in cases where the whole group was still connected in one cluster (or in a complete graph, as the mathematicians would say). In other words, there were no disconnected islands, and no isolated clusters. Yet, the group was no longer able to synchronize completely, try as they might. Too much information had been taken away.

One network was especially noteworthy, and that was the ring. Here we had arranged things so that people were connected in a circle (according to where they were sitting), meaning that they could only see their immediate neighbors (two LEDs) rather than all eight people. Recording the responses of all participants as a function of time on a computer gave us a lot of data to sift through. We found that although it was impossible for the group to come into any kind of persistent sync, correlations in button-pushing did show up between neighbors, and some correlation (although weaker) did persist even out to next nearest neighbors. Beyond that it went to zero, however.

Mass synchronization isn't limited to humans, of course, but is fairly ubiquitous in nature [68]. We see it in the coordinated motion of flocks of birds and school of fish. We see it in the flashing of fireflies of Southeast Asia where tens of thousands of male fireflies manage to flash in synchrony. Fireflies, really? Yes, indeed. In order to come into collective sync, no higher intelligence is necessary. But if not intelligence, then what else enabled collections of individuals to come together as one?

It turned out that if individuals just followed certain simple rules, then synchrony at the level of the system would often arise. For instance, for the fish in a school of fish, the simple rules might be to stay close to the fish around you (but not too close), line your direction up with that of your neighbors, and as Steven Strogatz tells it [69], get out of the way when a predator comes in. Computer simulations of *boids*, i.e. objects rigidly following these three rules, show behavior eerily familiar to us— swarming and flocking. The Boid algorithm is sufficient to realistically mimic such behavior. It is also a manifestation of emergence. No single fish knows what it is doing in the larger scheme. It has no concept of the cohesive motion of the swarm. There is also no command center—the system is entirely decentralized and distributed. Yet intricate collective motion nevertheless arises based solely on the local application of simple rules.

The same is true for the fireflies, crickets, and even neurons firing in sync. Although the biological details vary enormously, the global phenomenon of synchronized response is shared. Intrigued by the growing list of examples of mass-synchronization in biology, in the mid-1960s math-biologist Art Winfree set out to understand what the criteria were for synchrony. Could one come up with a generic mathematical model of biological oscillators and their interaction that would capture the effect of spontaneous synchronization? Winfree made a lot of progress in that direction, but it was a Japanese physicist Yoshiki Kuramoto who ultimately came closest to Albert Einstein's advice of keeping models "as simple as possible, but not simpler." In other words, make a model too complex and you won't be able to analyze it rigorously, make it too simple and it won't capture the important aspects of the observed phenomenon. Kuramoto had hit on the perfect balance in 1975.

What Kuramoto did was to simplify Winfree's original model by assuming that oscillators adjusted their speeds in a symmetric, pair-wise way depending only on their relative phases. So, the oscillator ahead in phase would reduce its speed and the one behind would increase its speed until they had come together. He was able to formulate a set of coupled differential equations describing this dynamics of the biological oscillators. Displaying brilliant mathematical analysis, he proceeded to work out an analytical solution to the simplified problem. This was a major accomplishment. Analytical, closed-form solutions to nonlinear differential equations are extremely rare, and here Kuramoto had done it with not just one differential equation, but N of them, where N could be any large number. Furthermore, he did not assume that all oscillators were identical—an unrealistic assumption in biology—but he allowed for a statistical spread in oscillator frequencies.

In the context of his generic model, what he proved was the existence of a phase transition. Synchrony within a set of fairly similar oscillators always occurred, but as the spread in speeds increased and oscillators became more diverse, there was a point where sync ceased altogether—a critical point. What's more, the precise location of the critical point came out of the theoretical model as well. The model was generic in that it didn't purport to describe any specific system correctly in detail, but it nevertheless captured something essential about synchronization and its appearance and disappearance. Refinements could always be made (and were) to fine-tune the equations and tailor them more to a specific application, but the model would always remain essentially "just" an *effective* model. It did not really have a chance of correctly describing the details of any system.

For that, one has to start with a specific system in mind and ask what the physical or biological variables are that describe the state of that system, and how these variables change based on the other variables and external forces. And even then, one really has only an approximate description of the system that works reasonably well on a particular scale. So in a sense, any model is an effective model relative to the scale for which it was designed.

That said, the Kuramoto model was a particularly glaring example of an effective model. Its formulation was not primarily driven so much by biological considerations as by mathematical concerns. Yet, at the same time, it was rich enough to show such counter-intuitive behavior as the presence of a phase transition. As we saw in Chap. 4, phase transitions hold special significance for the topic of emergence. Kuramoto had found a twist on the traditional phase transition lurking within ensembles of coupled oscillators; he had uncovered an abrupt global transition towards spontaneous temporal order.

Of course, biological sync can be more naturally appreciated beyond the mathematical abstraction. When we see the amazingly coherent motion of thousands of birds or fish, it is hard not to be moved. One has to have seen the phenomenon in real life, or at least watched a movie of it, to appreciate the magic. To us outside observers, it almost appears as if the collection of individual players had taken on a life of its own, as if the motion were that of a single organism. When a predator attacks, this is particularly startling to watch. Hundreds of starling eyes see the hawk coming, and the collective motion that ensues serves to minimize

Fig. 5.6 Murmurations of starlings: the rapidly shifting geometric patterns "look from a distance like a single pulsing, living organism" [70]. (Image by Walter Baxton, https://commons.wikimedia.org/wiki/File:A_wedge_of_starlings_-_geograph.org.uk_-_1069366.jpg, licensed under CC BY-SA 2.0)

the loss of starling lives. Sometimes, you even observe the dense group of individual starlings dynamically split apart as the hawk enters, and then reform after the hawk has left. It almost seems as if the cloud of birds had an intelligence of its own (Fig. 5.6).

Only that it doesn't. In this example of flocking or swarming, it merely seems that way. In fact, the collective cloud-like motion, the rapidly propagating waves, can be recreated on a computer by giving each "bird" a set of simple rules for its motion and a sufficiently fast reaction time. From these micro-rules of motion, the collective behavior then emerges. Here the term *emergence* signifies that the resultant global patterns of motion, the dynamics of the cloud if you will, are not entirely deducible from those micro-rules. Hence, the need for computer simulation—a kind of digital experiment, but not a rigorous mathematical prediction. On the spectrum between the longstanding sociological poles of 'structure' and 'agency', this example of emergence would be closer to agency. Social patterns are the collective effect that emerges from millions of individual actions. Adam Smith's *invisible hand* comes about as an unintended consequence of countless individual economic decisions.

In the last section we encountered the phenomenon of convection. There we came away with the notion that large-scale patterns, such as convection cells, can have their origin in effective theories insensitive to the microscopic details or composition. These global patterns get set up due to downward control, whereby the large-scale phenomenon restricts or constrains the motion at the small scale. It is an example more akin to 'structure' than 'agency.'

Both examples, convection and flocking, involve feedback between whole and part, but they highlight perhaps slightly different possible interpretations of emergence, or at least they embody conceptually different ways in which emergence can manifest in our world. What is the difference? It may not be a categorical difference, and more of a useful descriptive or conceptual nuance. Nevertheless, the natural starting point for the flocking phenomenon is the application of simple rules that operate at the level of the individual: "align your motion with that of your immediate neighbors," "maintain a certain fixed distance," and so on. When every member of the group applies these rules, larger structures can emerge (the cloud), which then acts back to modify the behavior of the individual member. How so? Not by changing the rules themselves, those are fixed. No, the feedback happens in a more subtle manner. It's in providing different concrete situations to which those rules get applied.

Let me illustrate using the bird example again. Since we now have a flock of swarming birds, by virtue of this larger structure, there are birds that are in the middle and other birds that are on the periphery. There are birds that are on the side facing the predator (a hawk), and birds that are on the opposite side. All of these birds will practically institute the same basic rules somewhat differently. Hence, the beautiful and dynamical motion of the flock itself. If every bird did exactly the same thing at the same time (as well as follow the same rules), the flock as a whole could never change shape. But that is not what is observed in swarming behavior. So while the rules of motion for individual birds are the same for every bird and static as such, the concrete implementation of these rules depend on the location of the bird in the context of the larger structure. In that way, there exists some weak downward control on individual birds. If it weren't for the larger structure, the very idea of "location" of the bird would lose its meaning.

The natural starting point to explain convection, in contrast, is at the macro level. Convection is a macroscopic phenomenon, and it can be understood only in the language of effective theories for fluids (the Navier-Stokes equations) that ignore the microscopic composition of that fluid. Here the notion of downward control appears to be stronger. Individual molecules are caught up and forced to participate in larger patterns not of their making.

We are left with two versions of emergence suggested by two different science examples, both postulating some form of downward control, but differing by degree and perspective. While swarming illustrates how individual behavior in aggregation can lead to unexpected and sophisticated higher-level structure, convection emphasizes how large-scale structures can act back on individual behavior. In many real-life examples, both feedback mechanisms operate simultaneously in a loop. Micro-behavior creates macro-patterns, which in turn influence micro-behavior.

Chapter 6
The Rise of Effective Theories

6.1 A Single Neuron

Let us revisit an experiment from the previous chapter. Say you wanted to understand why people pushing buttons can sometime synchronize given enough connectivity in the system. If you remember, in that experiment every participant was given the instruction to adjust their pushing of buttons in whatever way they wanted in order to come into synchrony with the others. We could ask: "What does it mean to understand this phenomenon?" Would we be satisfied to refer to the generic Kuramoto model? Probably not entirely—some of the assumptions are unrealistic in this setting. In the Kuramoto model, the oscillators are continuously tugging on each other's frequencies based on their respective phases. In our experiment here, people also adjusted their button pushing, but they could only do it in response to an LED flash on their terminal. OK, so what if we modified the model to reflect this pulse (as opposed to continuous) coupling between people's responses? Would we be satisfied now, if the refined model showed mathematical hints of synchrony?

Or is that still not realistic enough, and do we really need to go a level deeper? Should we be looking at more of the details of how light flashes get processed by people's nervous system? Perhaps we should be examining the phenomenon at the neuronal level, starting with what

© Springer International Publishing AG 2017
L.Q. English, *There Is No Theory of Everything*,
DOI 10.1007/978-3-319-59150-6_6

happens when photons strike the retina and excite the cones and rods lining it. The question I would like to explore here is whether the original phenomenon of synchronization would become clearer or more distant at that level of scrutiny. Would the explanation become crisper or more blurry?

Before we delve into this question any further, it is worth pointing out that there is no universal agreement among scientists. On the one hand, we have the reductionist optimism shared by some in the neuroscience community. For instance, a textbook written by three leading scientists, including Nobel Laureate in Medicine, Eric Kandel, contains the following lines in its preface [71]:

> behavior can be examined at the level of individual nerve cells. [...] Ultimately, these experiments will make it possible to study emotion, perception, learning, memory, and other cognitive processes on both a cellular and a molecular level.

On the other hand, we have the skepticism summarized by Nobel-laureate chemist Roald Hoffmann [7]:

> There are concepts in chemistry which are not reducible to physics. Or if they are so reduced, they lose much of what is interesting about them. I would ask the reader who is a chemist to think of ideas such as aromaticity, acidity and basicity, the concept of a functional group, or a substituent effect. Those concepts have a tendency to wilt at the edges as one tries to define them too closely.

Two Nobel Laureates, two very different takes. How can there be such fundamentally divergent views among scientists? Who is correct? If, as Hoffmann contends, some concepts in chemistry are irreducible to physics, then surely things like emotions and consciousness should be irreducible to molecular biology and chemistry. And surely, a thought or an emotion does not reside within, or is held by, a single neuron, let alone a single molecule. So what does it mean when neuroscientists say they study cognitive processes at the neuronal or even molecular level?

I think we can agree that the statement cannot be taken literally. Instead, the implication is that the neurons, being the parts that make up the brain (more or less), give rise to these collective, higher-level functions in a way that can be clearly delineated and traced back to them. In other words, something about these higher-level mental phenomena must already be contained in the way a neuron works. Stated this way, the assertion may sound quite reasonable.

However, as we have seen in previous chapters, the assumption that things can be *traced back* down, so to speak, has failed quite dramatically in many branches of physics, and in particular in condensed-matter physics. One could perhaps argue in defense of the neuroscientist's optimism that in physics what allows emergence to stand are the strange laws of quantum mechanics, which defy a strictly realist interpretation. In the world of neurons and macro-molecules, these laws do not exactly apply. (This, of course, could itself be seen as a manifestation of the emergent notion of 'protectorate.') Does an absence of quantum mechanical weirdness open the doorway for reduction?

One should still be skeptical. Many of our more recent examples of emergence did not invoke quantum physics. The lessons we learned in the realm of nonlinear dynamics of complex systems, of phase transitions and statistical mechanics, and so on, did not rely on quantum mechanics. What these examples had in common was the operation of some mechanism for downward causation.

Nevertheless, it is worth re-examining the issue of reduction and emergence in this new domain of neuroscience, and to focus again on the relationship between whole and parts in this context. In this examination it is wise to proceed by staying close to the empirical science and avoiding abstraction. In this way, we can more easily get a feel for the subject by following the level of complexity where-ever it leads us (and be it down the rabbit hole). I believe that in the process we gain an appreciation for the explosion of knowledge in this relatively new field of neuroscience and for some amazing work that has already been accomplished by people in this science. At the same time, we also come away with a clearer understanding of emergence and the limits of reductionism.

Back to our goal of understanding synchronized behavior in a group of people. As we saw, this group was easily and spontaneously able to come into sync as long as there were enough connections, but as the network became more dilute that ability disappeared. We could formulate a high-level *effective* theory or model to describe this experiment, perhaps starting with a refined Kuramoto model. This model would include several features we know are important, such as the participant's adjustment of their rate of button-pushing based on the majority (or, more precisely, some average phase) of the lights they see flashing in front of them. And it would include these features in a mathematically precise way, which is very likely to gloss over the details, but which allows us to make mathematically precise conclusions. One of those conclusions would show that indeed there is a critical point; as the communication network becomes more sparse, global synchronization disappears at a well-defined point.

Or we could reach for a more fundamental understanding beyond the power of this effective theory. We could try to understand neurons, neuronal pathways and networks. We might try to understand the structure and function of the visual cortex. From a reductionist perspective, it would seem logical to start with a single neuron, and then to work our way up to how neurons communicate with one another, all the way to how the temporal distribution of flashing lights are processed. So, having rejected the effective theory from before, and following instead the reductionist prescription, let's take the plunge. Ready?

A neuron is a single cell with a cell body containing the nucleus, a large number of dendrites, and an axon, as shown in Fig. 6.1. On this scale, then, the effective model of the neuron goes as follows. The dendrites receive signals from other neurons that connect to them. These signals are processed in terms of their combined amplitude, and if that amplitude exceeds a certain threshold, then an *action potential* is created that runs down the axon and terminates at the synapse. On the other side of the synaptic gap are the dendrites of other neurons. The action potential is a voltage spike of only about 1 ms in duration that is slightly asymmetric in shape with an amplitude (from base to peak) of about 80 mV. Due to the makeup of the cell membrane, this signal travels down the axon rapidly and reaches the synapse. At the synapse, it stimulates the release of neurotransmitters (contained in vesicles) that then traverse the synaptic gap to reach the dendrite of the

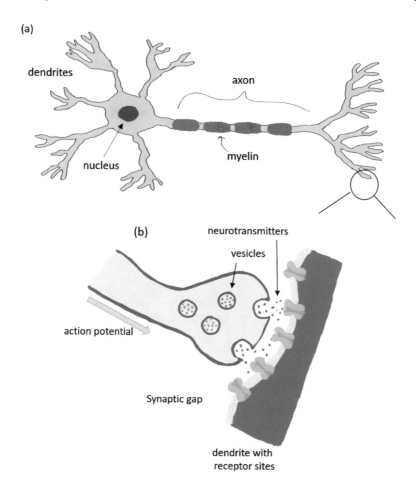

Fig. 6.1 (**a**) Schematic diagram of a neuron. The dendrites receive signals from other neurons, the axon transmits the neuron's action potential, and delivers it to the synapses. Also note the myelin sheath surrounding the axon. (**b**) Zooming in on the synapse reveals the vesicles containing neurotransmitters, which can get released into the synaptic gap

next neuron. At the dendrite of this next neuron, the neurotransmitters bind at receptor sites and cause an opening of ion channels. Thus a voltage signal is created that gets integrated with other incoming signals to determine whether this neuron too fires off an action potential.

This is the effective model of a neuron, or the quick and rough summary of it anyway. The generation of an action potential is how information is passed through the neuron, and it is the principal way in which neurons communicate and distribute information. But the model immediately raises many new questions. For instance, what happens at the synapse? We have no detailed answer at this scale; it remains mysterious. But even the question of why the action potential travels down the axon is not easily addressed? All we could say at this level of description would be that the neuron acts somewhat analogous to an electrical cable, where voltage pulses also propagate down a line. But how far does that analogy go? Surely, the neuron does not have a conducting copper wire at its center. Is its center conducting at all? Not exactly—the detailed mechanism for pulse propagation is quite different here.

6.2   Descending Down the Rabbit Hole

Following in the reductionist spirit, we have to descend further for answers. We need to reduce the scale and zoom in on the membrane of the neuron. This is logical because voltages (also called potential differences) are always taken between two points, and in this context they refer to differences between the inside and outside of the cell. It's the voltage across the cell membrane in other words. How can such voltages be generated across the membranes? In our everyday experience with voltage, these get produced by batteries or generators. How does the neuron do it? We need another effective theory at a lower scale.

It is the theory of ion concentrations and ion channels. Ions are atoms that are either missing an electron, such as potassium (K^+) or sodium (Na^+) ions, or have one electron too many, such as chlorine (Cl^-). These ions exist both on the inside and the outside of the neuron, but there are concentration gradients across the membrane. For instance, K^+ is more concentrated on the inside, and Na^+ more concentrated on the outside of the cell membrane. The combined effect of the ion gradients is that the voltage is negative on the inside relative to the

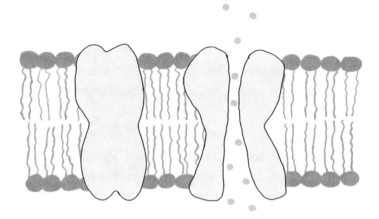

Fig. 6.2 The lipid bi-layer of the cell membrane, and the embedded ion channels that allow the influx and efflux of ions when open

outside. After all, where-ever there is charge separation (here in the from of ion separation), there are also voltages.

This is the resting voltage. But how is this voltage changed? Easy—ions have to flow across the membrane. But the membrane's lipid bi-layer is impervious to ions. That is true, but on closer inspection, every so often this bi-layer is interrupted by so-called *ion pumps* and *ion channel* which traverse the membrane; see Fig. 6.2.

We can visualize these ion channels as simple gaps in the membrane for now. There are two interesting facts about these channels. First, they can open and close in dependence on various factors, including the voltage. Secondly, they are specific to a particular ion. This second capability may be somewhat puzzling if we think of them simply as holes in the membrane. There seems to be something missing from the simple picture of Fig. 6.2. Take sodium and potassium, for instance. In the periodic table those two elements are in the same column, and the lighter sodium is directly above the heavier potassium. This means, not surprisingly, that the size, or ion radius, of potassium is larger. So, if the channels are just holes, wouldn't any potassium channel also allow sodium through, contrary to the observed result? If the hole is big enough for potassium, it is certainly big enough for the smaller sodium ion.

So, we realize that the hole picture is not the final answer, yet we could stop here, accept that these channels function a certain way and go from there. In fact, this is what many textbooks do; we do not necessarily have to know in detail how these ion channels work in order to appreciate how action potentials come about (more on this in a minute).

Or we could say, "No, my understanding is incomplete unless I first know how this ion selectivity can come about." Alright, in that case we keep going down. First, we have to recognize that the ions are not in a vacuum. They are surrounded by water molecules both in the cytoplasm inside the cell and in the inter-cellular plasma on the outside. Does that ring a bell? These ions are surrounded by a collection of water molecules that are attracted to their charge; they are not "bare." It's funny how similar concepts can creep up in vastly different areas of science and at very different scales.

The result is to increase the effective radius of these ions. Since the bare Na^+ ions have smaller radius but the same charge, they attract the water even more strongly, and so their effective radius turns out to be larger than that of the K^+ ions. Thus, the ordering in size is reversed by the presence of the cloud of water molecules. Still, the conundrum doesn't yet go away. How can there now be sodium specific channels? Shouldn't anything that lets the larger sodium (plus water) through also accommodate the potassium (plus water)? And also, how can they be "voltage-gated", in other words how do they open and close in response to a membrane voltage?

For that, we need to go down in scale even more and examine a single ion channel in detail. It turns out that the ion channels are by no means simple holes in the membrane. No, they are complicated macro-molecules. A schematic of one is shown in Fig. 6.3. What happens near the top at the selectivity filter is that the water molecules surrounding the potassium are first stripped off. How so? Well, we are entering the world of quantum chemistry, and so the explanation will be less mechanical and more energetic. It turns out that the atomic tips of the filter bind to the ion more strongly than the water molecules do, and so it becomes energetically favorable to break the water bonds

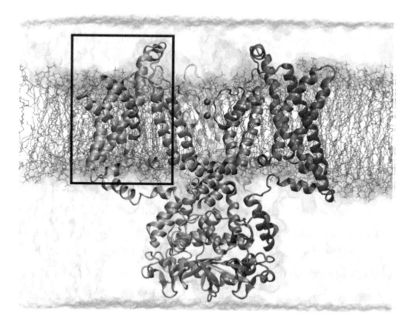

Fig. 6.3 The potassium channel on a molecular level. The ion selection happens at the top, where the carbonyl oxygen atoms from each of the four helical subgroups create an opening or pore for a bare potassium ion to traverse. (Image from [72])

and temporarily become attached to the filter. This attachment is weak enough to eventually allow the bare ion to pass through where it can hook up with water molecules again.

So why can sodium not do the same? Why not also shed its water entourage and bind to the filter? The simple answer is that the energetics don't favor this path. The sodium binds more strongly to its water molecules, and its bare ion-radius is smaller, which means that it can't bind as strongly to the filter atoms, the oxygens. Thus, the sodium would rather keep its water entourage and thus can't pass through.

Starting from the neuron as a single cell, we have already gone down in scale twice: first, when we zoomed into the cell membrane and noticed ion channels, and second, when we homed in on a single ion channel and discerned its molecular structure. However, from the perspective of physics, even this magnified scale of the potassium-channel protein is incredibly large and complex. The protein contains

four helix sub-units which are arranged to form a narrowing funnel
shape (as indicated in Fig. 6.3). All in all, we have tens of thousands of
atoms in each of these channel proteins. There is still a long way down
to where we would analyze things using the Schrödinger equation, or
any tool of quantum mechanics for that matter.

The point is we could easily keep going down another few rounds.
The question, though, is if this path towards molecular chemistry and
quantum physics really elucidates why action potentials travel down
the axon? That was one of our original questions, after all. At these
small scales, however, we don't see the neuron as a functional unit. The
neuron, the axon, the membrane, they all dissolve into assemblies of
macro-molecules and hydrogen bonds. The phenomena associated with
the larger scale also vanish from view. Instead of a propagating action
potential, we only witness single ions making it through the ion channel
at certain rates by shedding their water-companions.

While this exploration of the ever smaller parts was certainly fascinating and informative, and the insights learned there valid, practically
speaking we have to re-emerge and head back up in scale. Otherwise,
the action potential will recede further from view. We will not encounter
it on our journey deeper into the microcosm, let alone capture any
higher-level cognitive task there. We have to zoom back out. At the
smaller scales we get answers to questions pertinent to that scale. So
we learn about how ion channels function to select for different ions,
for instance, which is a question that can only be addressed at this
level. There is no doubt that answers to such questions are immensely
useful, for instance, in designing certain drugs that act on that same
bio-molecular level.

More broadly, however, this insight into the microscopic function of
ion channels does not magically solve our original problem at the larger
scale of signal propagation down the axon. We will need two different effective theories to describe the two phenomena of ion-channel
selectivity and action-potential propagation. Granted, these two models
can make contact with and inform each other at various junctions.
Nonetheless, there is no single theory with a unified description that can

capture both phenomena simultaneously. Moreover, this is true in spite of the fact that these two phenomena are actually pretty close to one another in the grand scheme of things; they are not so different either in scale or scientific context.

6.3 Re-Emerging from the Rabbit Hole

Clearly what we need to do is to zoom out, and that is exactly what neuroscientists do when they want to understand the propagation of action potentials. They leave the microscopic details behind and accept that things operate a certain way without at that time investigating exactly how. And, ironically, they now build very generic models based on the rules and language of electronic circuits. Let me illustrate this point, since it is quite revealing of the actual scientific process, but rarely ever explicitly discussed.

As we just saw, the cell membrane can separate charges, when the active ion channels are closed. What else does that? If you know a little bit about electrical circuits, capacitors come to mind. A capacitor is an idealized circuit element that holds charge on its two conductive plates. So we can model the membrane as a capacitor—they are similar in the sense that they both separate opposite charge across a dielectric (non-conductive) material. But why stop there in our search for circuit analogies? Since some ion conductance through the membrane does happen, we can model that aspect by a resistor—a circuit element that allows a current to flow through it when there is a voltage across it. Now the phenomenon of opening and closing of ion channels can be represented by a variable resistor. The ion gradients (sodium, potassium and chlorine), moreover, can be modeled as batteries. After all, differences in ion concentration between outside and inside of the neuron creates a drive similar to what a battery would do. (This statement can be made more rigorous by considering a voltage that would exactly balance out the effect of the diffusion of ions due to the concentration gradient, leading to the so-called Nernst voltage.)

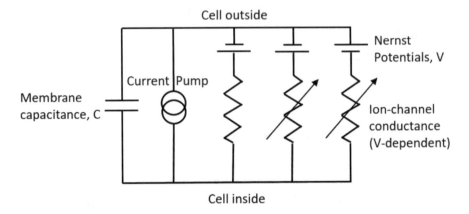

Fig. 6.4 A single Hodgkin-Huxley cell comprised of variable resistors, constant current sources, a capacitor and batteries modeling the ion flow across the neuron membrane. To get to the full axon, imagine these cells connected via resistors that transmit local voltage swings to neighboring parts of the axon

So now we have a model with resistors, batteries and capacitors—see Fig. 6.4. These equivalent circuit diagrams were first worked out by Hodgkin and Huxley in 1952, and based on them they were able to match the measured shape of the action potential exactly and also predict a correct propagation speed down the axon. This was at a time when the details of the ion channels were not understood. Certainly the molecular structure of the proteins were completely unknown at that point. It wouldn't be until the late 1990s for those to become known through X-ray crystallography—more than 40 years after the model was proposed.

In the Hodgkin-Huxley model, the speed of propagation is now treated as it would for an electrical cable—it is formulated in terms of capacitances and resistances. For instance, if the capacitance of the membrane could somehow be reduced, then the signal would travel faster. Indeed, nature does that in the form of wrapping an insulating layer (myelin) around the axon at certain places, decreasing the local capacitance. This gives us further confidence that the effective model captures key features of reality, even though it doesn't make reference to biological processes at all and is, in fact, formulated in the language of electronics! It is a remarkable state of affairs, if you think about it.

Once we have our phenomenological circuit model, we understand the transmission of the action potential on that level, and it is no longer necessary to know the neurophysiological details underlying it. We can practically ignore the fact that the action potential is created by an opening of the Na^+ channels which allows the influx of sodium ions into the cell, making its voltage more positive and *depolarizing* it; that this opens up even more of the voltage-triggered channels, leading to further influx until the voltage reaches the sodium Nernst potential at the peak of the action potential. We don't need to know that now these channels begin to close and the potassium channels instead open, leading to a outward flux of K^+ ions, which eventually restores the negative potential inside the neuron. We don't have to worry about how this massive but local buildup of sodium ions inside the cytoplasm affect neighboring parts of the axon, namely that sodium ions can diffuse and migrate to those parts, and so can potassium ions on the outside. Nor do we need to appreciate that this causes those places also to depolarize and to start the process of opening channels. In short, we don't have to understand how the action potential is able to move in space along the axon on a molecular level.

Of course, ultimately this is the correct biological explanation. On the whole, the circuit model is generically consistent with these more detailed mechanisms, but it also doesn't rely on them exclusively. There are many possible biological processes that could all be described by this generic model. The conductance across the membrane doesn't have to be due to ion channels, and those ion channels certainly do not have to be made from those specific proteins. The model is completely silent and neutral on those details of the substrate.

So why do we need the generic model then? Why can't we just do away with the circuit model and focus only on the biological mechanisms? That's the important question whose answer has larger implications. Intuitively, it seems that the more we know about and stress the correct microscopic details, the better positioned we should be for understanding the cell as a whole. This example reveals, however, that sometimes the better strategy is to block out the microscopic substrate and construct effective models.

Questions like "How fast is the action potential predicted to move?" are only answerable by moving to something like the circuit model. The reason is that it can be translated into mathematical equations, and these can be solved numerically. In fact, we can do even better: what now becomes possible is to further modify and simplify the mathematical description of the problem, while preserving the main features, until it yields to an analytical solution. At this point, we finally crafted a model that allows us to prove that action potentials will form given certain initial conditions and that they will propagate at a speed that can be predicted not only by numerical simulation, but by mathematical analysis. Many more exciting insights can be gleaned from this kind of approach [73, 74]. Our ability to prove things abstractly using mathematics has come at the cost of divorcing the model from its original biological context, of course. But this divorce from context, this process of abstraction, was necessary in order to arrive at a model that could be fruitfully analyzed using mathematical tools. This kind of divorce happens all the time in science, and it precludes the possibility of one grand unified theory of everything.

Where does this leave us? Well, we are now at the stage where we have constructed an effective theory of the action potential using biological insights as a guide. This theory glosses over the finer microscopic, bio-chemical details, but works very well at the scale of single neurons. However, our original goal was to learn something about visual and cognitive processes to address our original question about synchronization within a group of people. We are obviously still very far away.

In the hypothetical journey back up the rabbit hole towards that goal, we would next investigate how neurons communicate in more detail, and how those inter-neuron connections get strengthened and weakened; we would have to examine the neural network and its plasticity. On this journey, we would get diverted at certain junctions back down in scale. For instance, we would need to understand how the chemical transmission of information between neurons works, which would lead us into the bio-chemistry of neurotransmitters. We would need to look at feedback mechanisms that modulate the connectivity between sets of neurons.

There can be no doubt that this is all extremely interesting science. But it is doubtful that neurotransmitters and synaptic receptors will inform our understanding of human synchronization in any way. The gap between ion channels and the axon's transmission of voltage pulses was actually fairly small, yet we saw that the effective theory already glossed over the details. Now the gap in both scale and complexity between neurotransmitters and high-level cognitive behavior is clearly vastly larger. Any scientific details on the small scale would long be washed out by the time we arrived at the larger scale.

In a sense, this should not come as a surprise. Let's recall what is actually involved in the task given to our participants. They needed to synchronize their pushing of a button to the lights flashing before them on a panel. It seems like a simple task compared to most others we have to accomplish in our daily life. It seems almost trivial in comparison to driving our car to work, completing the projects and assignments when we get there, carrying on a conversation with co-workers, navigating the social expectations, and so on.

Yet even this simple task, when deconstructed, does not seem so simple. It assumes that we have processed the instructions and understand the concept of the temporal ordering of events. It assumes that we can become aware of a set of blinking lights and their times of occurrence. It assumes that we can somehow integrate these times and compare the average time to that of our own pressing of the button. It assumes that we can make a change in timing of our response based on this information.

It becomes clear that this task requires the coordinated participation of many millions of neurons, located in many different and far-separated parts of the brain, including in the visual cortex, the primary motor cortex, the temporal lobe, and so on. On the level of single neurons, implementation of the task is intractably complex. There is little hope to approach the phenomenon at that level, as it doesn't really exist on that scale. In no way does a single neuron, or even a collection of neurons, hold the abstract notion of synchronization and synchronized response. Neuroscientists, of course, know this quite well. They look for functional units within the brain, sections comprised of large numbers of neurons that can act in concert to perform a certain task. Often such a unit is characterized by its clustered connectivity.

The upshot is that, if we want to have any chance of understanding the collective phenomenon of human sync (which in the final analysis involves a group of people), we will have to start at a larger scale. Many layers of effective theories will lie between that higher-level description and single neuron processes. At each layer some abstraction and also some simplification will occur—some sort of course-graining, if you will. It is necessary to lose the microscopic details and to zoom out, in order to grasp things at a higher level.

Neurobiologist William Newsome cast an apt analogy [66] to underscore this idea: If the object is to learn how a software package, like Microsoft Word, functions, it would not be an even remotely viable strategy to examine the microelectronic hardware, the turning on and off of transistors, one at a time. Even though the software ultimately manipulates bits which get encoded in flip-flops which in turn are made up of many transistors, it is not fruitful to start there. The program has to be understood at higher levels of organization. It has to be approached perhaps at the level of the software code, or, more likely for most of us, experientially at the level of producing text by starting to type on the keyboard and seeing what happens on the monitor. The fact that the program can work on a number of platforms with different processor architectures should tell us that the phenomenon we are interested in (the piece of software) does not rely on a particular microelectronic structure. There is no one-to-one relationship between the microscopic structure or processes and the higher-level behavior.

We recognize here a version of the "complexity obstacle" to a reductionist understanding of cognition. One master theory cannot explain the diverse complexities that arise and compound within each new layer. The only successful strategy is to employ a cascade of effective theories—each responsible for its own phenomenological layer. The fact is that while these effective theories do "talk" to one another, they cannot be fully merged. They were constructed that way. The Hodgkin-Huxley circuit model of the neuron is an abstraction and simplification, but that is precisely what allows for mathematical prediction. The biochemistry from which the model emerges is qualitatively different in description.

What sometimes confuses the issue is the role of mathematics. Models are increasingly becoming quantitative at any level, which means that mathematics is now involved on many fronts. Whereas, in the past, fundamental physics had long been the lone science closely associated with mathematics, more recently the biological sciences have increasingly moved in that direction as well. The ubiquity of mathematics may misleadingly suggest that a deeper unification had been accomplished. Coming back to our example of the neuron, we see mathematics employed in circuit models of the axon, but we may also see it in the context of diffusion and reaction (or binding) rates at the molecular level. We may see it again in numerical models of the configurational dynamics of proteins. This does not imply, however, that any type of conceptual convergence has taken place. It simply means that models have become more quantitative at all levels of description.

Emergence in the realm of scientific models thus represents another line of attack on reductionism—one that takes center stage when we contemplate the success of effective theories and what it means to understand a phenomenon. But within the field of neuroscience there is an additional line of reasoning that we can bring to bear. As we have seen, one formidable obstacle to the reductionist project is posed by downward causation or downward control. We have already encountered the notion of downward causation in our discussion of pattern formation in nature, and it is no surprise that it would play a major role here as well.

What MRI scans clearly show us is that there are complex and large-sale patterns associated with various cognitive tasks. Certain groups of spatial regions of the brain "light up" when we think certain things. This correlation between mental activity and brain patterns as captured by MRI is, of course, an enormously useful tool in medicine and neuroscience. It has even allowed us to communicate indirectly with people suffering from locked-in syndrome where they are completely unable to physically move any part of their body. It is not hard to imagine technology becoming commonplace in the near future capable of responding to mental cues only.

So the formation of dynamic patterns in brain activity, involving many millions of individual neurons, seems to be a key aspect to our cognition. Individual neurons get swept up in these larger emergent patterns of neural activity. What causes these extended brain patterns to arise in the first place? On the level of the physical brain, the answer must often be: a previous brain pattern. Patterns beget other patterns, and so on. This implies that the evolution of the system will not be able to be understood at the level of the single neuron, but must be approached at the level of larger-scale self-organization. The single neuron is likely as "trapped" by the larger neural patterns as a water molecule is by the convection cell in which it is caught up. Or, to recall a lesson from chaos theory and the butterfly effect, in complex systems it is generally impossible to say which set of local events caused a particular global phenomenon. Translated into this context, no single neuron in isolation could ever be identified as having given rise to or caused any particular brain activity. Things do not come down to the firings of a single neuron.

Our journey has taken us from quantum mechanics to statistical and condensed-matter physics, from dynamical systems and complexity theory to neuroscience. Conceptual breakthroughs within all these far-flung branches of science point towards a common philosophical insight that has arguably been under-appreciated so far in the broader discussion. They point to emergence as an important guiding principle in much of contemporary science. The common thread is that such things as structure, network connectivity, self-organization, feedback, and renormalization are central in explaining some of the most interesting real-world phenomena. Furthermore, these examples from contemporary science have helped define and circumscribe what we mean by emergence more sharply—a term often used loosely in the broader literature. Having thus demonstrated the limits of reductionism in the physical sciences, as well as in the biological sciences, the next chapters of this book aim to illuminate the importance of emergence more broadly—beyond the boundaries of science. In the words of Phillip Anderson,

> The arrogance of the particle physicist and his intensive research may be behind us, but we have yet to recover from that of some molecular biologists, who seem determined to try to reduce everything about the human organism to 'only' chemistry, from the common cold to all mental disease to the religious instinct. [4]

Chapter 7
Social Emergence

7.1 Social Networks and Normative Behavior

When I tell my students that I didn't know what an email was until my sophomore year in college they seem genuinely shocked. I like to think it's partly because I don't really look very old. More likely, though, it's because they have grown up all their lives with social media of various kinds that it actually strains their imagination to think what life would be like without all of that. How would you get information on stuff without Google and Wikipedia? How would you stay in touch with friends without cell phones, Skype, Instagram, and Snapchat? Even for me it is now sometimes hard to believe that I hadn't always possessed those things.

Recently, I was pondering giving a lecture on social networks to a bunch of my freshmen, and it quickly became clear to me that it probably wouldn't go over too well. After all, this is not an academic subject to them, this is their everyday experience—it's as natural as breathing. So my next thought was, why not exploit that. Instead of a lecture, what I should do is to design some kind of group activity where I could set them loose exploring their own social networks online. It turns out the college's site license to a program called Mathematica made it possible for us to do exactly that. All that was necessary was to allow Mathematica to import one's Facebook contact information, and it would then do a whole bunch of quantitative analysis on that data.

When I tried it out on my own Facebook data, the results were instantly intriguing. There in front of me on the computer screen was my social network of Facebook friends visualized in the form of a giant graph. The nodes of this graph were my friends and the lines running between various pairs of nodes were the friendship links. But what's more, Mathematica had figured out in an instant how to arrange all the nodes on the screen so as to make the graph as visually clear and simple as possible. All it needed was the names of my friends and their mutual friendship status—nothing more. Yet, the algorithms had somehow uncovered the deeper structures of my network. It had magically put all my family members in one corner, all my college friends in another, and all my work colleagues in yet another. It had even found a way to distinguish between my colleagues and students at Dickinson College. In other words, it recognized the distinct communities that made up my social network on Facebook. And it did so just by sorting through tons of network connections.

Just visually scanning the graphical representation of the network, you could actually learn quite a bit. The various communities within my social network were clearly characterized by lots of dense interlinking. Two friends within such a community were very likely friends of each other as well. The technical term for this phenomenon is *clustering*. There were some isolated clusters which that made perfect sense. I had been guest lecturer at a SUNY one semester, for example, and accumulated some friends that way. But the graph also revealed people that belonged to two of my communities simultaneously, and by this virtue alone they formed a bridge between what would otherwise have been separate islands. So, for instance, I have a colleague in the Chemistry department who also happened to go to the same college as I did. Not surprisingly, she and I share some of the same friends in either community. Within my global network, she represents a sort of linkage between distinct clusters (Fig. 7.1).

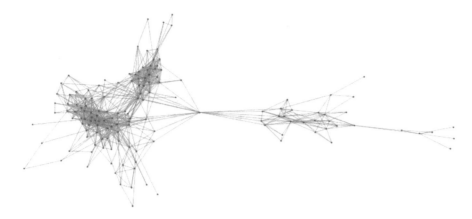

Fig. 7.1 Part of my social network of Facebook friends visualized

It is these links between clusters that make our world "small." We can get to almost anyone in the world in just a few short steps—hence the proverbial six degrees of separation. And this fact, as Duncan Watts emphasizes [75], has profound implications for everything from the global dissemination of ideas to the spread of diseases. The structure of social networks determines to a large extent how they function. Even in my small network of Facebook friends, the connectivity structure clearly reflected real aspects of my social life. The phrase coined by American modernist architect Louis Sullivan that "form follows function" seems to apply here as well. Structure does seem to mimic function—and vice versa.

Individual autonomy versus collective reality; freedom versus destiny; agency versus structure. The field of sociology has long wrestled with the recursive relationships between individuals and the larger social networks to which they belong. Individuals, of course, make up the larger society, and individual actions, attitudes and choices contribute to societal norms, institutions and change. Conversely, society shapes and circumscribes individual actions. We recognize the same recurring theme of feedback between whole and parts, this time playing out in the very different context of social networks.

But what really are social institutions and how should we understand social structure? These types of ontological questions have occupied sociologists since the beginning of their field going back to the mid-nineteenth century. In fact, sociology was one of the very first fields of academic study that envisioned and formulated notions of emergence. It simply seemed hopelessly inadequate to understand things like social organization and stratification, or institutional and cultural idiosyncrasies in a purely reductive manner. Reductionism was the program of the other sciences at the time, but it didn't seem to fit in this arena. For one, the sheer complexity of social processes and conditions seemed to make this approach untenable. And secondly, social institutions did not seem to be reducible to individuals in a plausible or natural way. To the contrary, they seemed to guide and constrain people's lives. No experiments could be conducted that would somehow simplify this phenomenological richness found within the fabric of human society; one had to face it head on and all at once. It's not like one could simply find some complete loners out there who grew up in a complete social vacuum from birth.

Hence, the structuralist view came to dominate sociology early on through the work of scholars like Emile Durkheim, Karl Marx, Max Weber and others. As the name implies, this view emphasized the role of larger-scale structures—institutional, cultural, or political—in shaping individual behavior. A second generation of sociologists, lead by Talcott Parsons, built on these ideas and tried to put them on a firmer theoretical foundation. For Parsons, the role of social norms were particularly pivotal in shaping and defining institutions. By this time, notions of social emergence were formulated quite explicitly and in a way that we would immediately recognize:

> It is not possible simply to "extrapolate" from the personality mechanisms of the one to those of the many as participants in the social system. [76, p. 45]

The quote, of course, expresses a key feature of emergence, namely that of genuine novelty found at higher levels of complexity or organization. You can't look at individual "personalities" and derive from them the character of a social group, if you will. Indeed, even the idea of self-consistency that we have repeatedly stressed in our discussion of feedback between levels is clearly stated in the sociological literature of that era. A little earlier in the same book by Parsons, for instance, we read that:

> A social system cannot be so structured as to be radically incompatible with the conditions of functioning of its component individual actors [. . .] [76, p. 27]

After all, a social system is comprised of 'individual actors' and therefore cannot ask these actors to do something that they cannot do. Without this self-consistency in place, the social system cannot conceivably exist at all. So how does the feedback actually work, and what mechanisms are operating underneath it all?

A book by sociologist Dave Elder-Vaas provides some insightful answers [77]. The author first argues that one of the elementary emergent entities related to social structure are *normative circles*. A normative circle, or norm circle, is basically a group of people that are bound by a set of shared norms. What makes you a member of a norm circle? You agree to act according to those norms, or codes of behavior, you expect others in that group to uphold those norms, and you may enforce the norm by sanctioning any transgression of it by other members of the circle. Since members have the expectation to be somehow reprimanded if they violate the norm of behavior when in the presence of other members, it psychologically reinforces the norm which then modifies their behavior.

The important point is that the mechanism of downward control is mainly psychological in nature. A mental expectation is instilled in members that compliance with the norm will elicit positive responses by other members, perhaps in the form of their approbation, friendly demeanor, or some other reward, whereas violation of the norm will meet with disapproval or sanction. In this way, members of the norm circle will tend to comply with the set of norms even if they don't individually endorse or agree with them.

So a normative circle is more than just a collection of people who happen to share a common belief or adherence to a norm. Rather, the norm is psychologically reinforced by the mere fact that one is a member of the circle. In fact, the membership by itself can override one's own personal beliefs when the two are not aligned. Colloquially, that invisible (but experientially real) force is referred to as "peer pressure." One example of a norm circle would be members of a gym, where certain norms about treating other gym members and the equipment are internalized by all and do not have to be explicitly stated. Or think of a college sorority or fraternity. By joining a student implicitly agrees to follow certain rules, codes of conduct, or ways of acting. Some of these rules may only reveal themselves subsequently by trial and error and by the social feedback one experiences as a consequence. Here, the circle is fairly small and contained, and so it is pretty clear when one is in the presence of another member and when one isn't. And in social situations where one is not in the presence of other members, the norms do not necessarily apply any-longer; one does not necessarily feel bound by them. So during summer break, college students typically leave campus and go back home to their families, and they thereby enter a different norm circle with different sets of expectations and constraints on their behavior.

In other cases, the boundaries of the norm circle are not entirely clear to any one member. This usually happens with larger and more amorphous norm circles. In these situations, one cannot possibly meet every member of the norm circle in person, and so from the individual's perspective, the circle becomes an *imagined norm circle* [77, Chap. 6]. We can only guess at and impute the size and shape of it. As a result, our actions are influenced not so much by the actual norm circle, but by our conception of it. We make the mental calculation that whenever we are around people of this particular background or in that particular context, we better observe the given norm, but not necessarily at other times.

This is a curious state of affairs when you think about it. If we were omniscient, we could know exactly its individual composition. But we aren't, and so we find out gradually about the existence of the norm by people repeatedly chiding us in public for a certain behavior in contradiction with it. From these consistent individual interactions, we start to get an inkling that this can't be a random coincidence and that instead we are confronted with a societal norm enforced by members of this norm circle. But we can only vaguely make out the extent of this circle. It has a certain depersonalized character to it. In these situations, it's almost as if we come face to face with societal phenomena channeled through our interactions with individuals. On the largest scale, the 'nation' could perhaps be seen as an emergent phenomenon that arises from the accumulated extrapolations of many individuals about the extent of a particular set of norms. These extrapolations then act back on the actual norm circle, creating a certain manifest, albeit constructed, reality.

7.2 The Wealth of Nations

A good first illustration of the dramatic effects of social feedback from collective ideas of nationhood is Korea. Before its division in 1945, of course, Korea was one country with a common language, culture, literature, history, religion and so on, all of which had developed over the span of centuries. In fact, political unification of most of the Korean peninsula was achieved as early as 676 AD. At around the same time, Buddhism became a large part of the culture throughout Korea. A second, more durable political unification was instituted with the arrival of the Goryeo dynasty in 918 which would last through conflict with Mongols until 1392. The next dynasty, Joseon, established a strong central bureaucracy and managed a number of political, social and economical reforms that cemented its rule, and it saw the wide adoption of the Korean alphabet and the spread of Confucian philosophy. The Joseon dynasty would last for an incredible 500 years, surviving various conflicts with, and occasional invasions of, neighboring powers and confronted in the later period of the dynasty with Western imperialism.

So even a cursory examination of Korean history shows that while it found itself, at times, within the sphere of cultural and political influence of China and Japan, Korea developed strong central institutions that then made possible, or actively enabled, the wide dissemination of social norms, cultural practices, and religious ideas. It remained ethnically homogenous throughout its history with few instances of mass-immigration or foreign settlement.

In 1945, however, a severe bifurcation occurred whereby the country was split apart along the 38th parallel. North Korea, coming into the sphere of the Soviet Union immediately after the Second World War and then China, saw the establishment of a Communist dictatorship, whereas the South developed democratic institutions and a capitalist economic system rooted in principles of free enterprise, open markets, and connected to international trade. North Korea evolved into one of the most secretive and isolationist regimes on the globe with virtually no contact to the rest of the world (Fig. 7.2).

Fast forward to the present. Korea has now existed as two separate states for more than 70 years. Statistics from the year 2011 indicate that the per-capita GDP of North Koreans was $1800 [78]. Life expectancy is slightly below 70 years. Wide-spread food shortages being prevalent, North Korea relies on food assistance from China and from international relief organizations. The absence of such assistance between 1995 and 1998 lead to mass starvation and famine; during this time it is estimated that around 2.5 million people starved to death [80, 81]. Things have improved since, but even now, roughly a fifth of children 5 years of age and younger are estimated to be underweight, and the infant mortality rate hovers around 25 (per 100,000 births) [78]. Most everyone lacks access to a car or motorcycle, the population is almost completely cut off from the global internet and from international news and entertainment. Most North Koreans are unfamiliar with the kinds of consumer electronics (such as computers, laptops, smart phones, MP3 players) or electrical appliances taken for granted in the developed world.

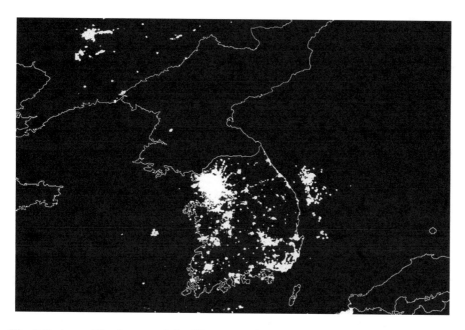

Fig. 7.2 A satellite image of the Korean peninsula at night. North Korea appears almost entirely dark. (Image by Roman Harak; https://commons.wikimedia.org/wiki/File:North_Korea_-_Satellite_view_(5015891270).jpg; licensed under CC BY-SA 2.0)

South Korea meanwhile boasts a thriving, diversified and modern economy with innovative, multi-national companies that design, fabricate and export exactly those modern consumer products and appliances that are unfamiliar to the average North Korean. In fact, South Korea is now internationally known for its leadership in consumer electronics products, such as smart phones and tablet PCs. The per-capita GDP is around $32,800 [78], about 18 times that of North Korea. South Korea has become a major producer, as well as exporter, of automobiles that can compete on the global market. The infant mortality rate is down at 4 (per 100,000), and the life expectancy of an average South Korean is about 80 years, or 10 years longer than his or her North Korean counterpart.

The raw statistics, of course, cannot begin to capture other vast quality-of-life differences experienced by people in the two Koreas.

South Koreans take for granted basic civil and political rights. There are strong political parties, a free press and uncensored access to information, independent courts, labor unions representing the interest of workers, security for property owners, patent rights for inventors, the list goes on. A strong middle class has emerged with substantial purchasing power and economic freedom. In this overall climate, not surprisingly, the arts and the sciences have been able to flourish, as has entertainment and commerce.

It is fair to say that if we picked two people at random—one from Seoul and one from Pyongyang—their life experiences, their beliefs and outlook, their health, their education would all appear highly divergent. Mi-jung from Seoul would have surprisingly little in common with Jae-eun from Pyongyang. In their outward appearance, Mi-jung and Jae-eun would be easily distinguishable, certainly in terms of clothing (jeans are very popular in the South but forbidden in the North), hair-style, and mannerisms of speech and gesture, but most likely also in terms of weight and even height. Many refugees from the North who manage against the odds to flee to the South find out that it is difficult for them to assimilate in their new-found home. Even the language in the South has evolved in ways that can make it hard for a recently escaped North Korean to follow [80], incorporating many English phrases, for instance. And then there is the cultural shock and disorientation of seeing a modern metropolis like Seoul with its skyscrapers, its whirlwind of traffic and construction, its variety of stores, shops, bars and street vendors, its dizzying neon signs and billboards. Add to this the unfamiliar cut-throat competition for jobs, housing, and social status and acceptance. As one North Korean escapee put it: "The difference between North and South is like jumping ahead a century." [80] In fact, Mi-jung would probably have more in common with Emiko of Tokyo, Japan—a former military adversary and conquering power—than with Jae-eun. Why is that?

What makes the Korean example so poignant is that the answer is fairly straightforward, self-evident even: it has to do with larger-scale social structure, with systems of government and economic organization. From a detached academic perspective, Korea is thus the perfect historical experiment—one of the cleanest test cases handed to sociologists in the laboratory of world history. Think about it: the initial conditions are set up to be extremely similar, then by historical accident an arbitrary line is drawn, as a result two societies form and are allowed to evolve according to very different rules and with very little contact to one another. Seventy years later, the two countries are nothing alike, they are dissimilar beyond recognition. The inhabitants of the two countries are separated by much more than an impenetrable geographical border.

Mi-jung and Jae-eun are both ethnic Korean; any dissimilarities between their respective countries cannot be attributed to ethnic, racial or genetic differences. And historically, those dissimilarities didn't exist. If you took a random resident from Seoul and one from Pyongyang a hundred years ago, no deep systematic differences would have shown up. Yes, there always existed some regional variations in dialect, custom, cuisine, and so on, but those differences were shallow compared to the deep unifying cultural bonds. And yes, there were geographic differences, idiosyncrasies in the lay of the land, the abundance or scarcity of certain natural resources, but those variations exist everywhere in the world and usually don't interfere with the cohesion of nations. Even Jared Diamond, whose writings (and in particular his first book [82]) is sometimes misinterpreted as advocating for a rigid kind of geographic determinism, has rejected such attempts at explanation. Instead, it is intuitively clear that the manifest difference in the fates of the two countries is entirely the product of the different societal conditions, brought about by different organizational principles of government and commerce, into which Mi-jung and Jae-eun were born and grew up. The difference can not be attributable to initial conditions. If anything, the northern regions of Korea used to be somewhat more prosperous before 1945.

Nor can one argue that the differences are temporary transients that will die down and go away in the course of time. No one can credibly claim that in the end, the two countries will asymptotically converge, as it is the genetic makeup of the population that primarily determines a nation's trajectory. Even if such explanations have fallen out of favor in academia, one could argue that they are still, in one form or another, quite prevalent in the popular imagination. In the Korean example, however, it is very clear that no such convergence has happened or is about to happen naturally (in the absence of external intervention). The political institutions and structures on both sides of the border have found ways to stabilize and perpetuate themselves within society. In the North, the institutional entrenchment was achieved as always in totalitarian regimes—through tight, top-down control of resources coupled with a harsh security apparatus. Here the feedback is primarily one-way and characterized by state intimidation of the population. In the South, institutions also exercise forms of control over the lives of individuals, and law-enforcement also exists to discourage violation of societal rules. The crucial distinction is that, in principle at least, these forms of control exist with the 'consent of the governed,' and that they abide by (and are constrained by) legal principles. The constant calibration and adjustment that occurs in a democracy between individual liberties and societal needs produces institutions that have been organically integrated into, rather than thrust upon, the larger society.

In the book "Why Nations Fail" [83], Daron Acemoglu and James Robinson lay out an institutional theory of poverty and prosperity. A physicist might be tempted to call the arguments put forth in this book the *unified theory* of world poverty. The scope of their theory is broadly encompassing, ranging from the regional to the inter-continental distributions of wealth. Why, for instance, is South America quite a bit poorer than North America, Western Europe richer than Eastern Europe, and Botswana so much richer than its regional neighbors?

The authors reject as empirically inadequate or contradictory a number of traditional and widespread explanatory hypotheses, including the geographic hypothesis, the cultural hypothesis, and the ignorance hypothesis. The cultural hypothesis, for example, holds that certain cultural attitudes prevalent in certain regions of the globe prevent the accumulation of wealth. A weak work ethic or distrust of other people or institutions are often cited. But such cultural arguments, besides usually tainted by prejudice and thus intrinsically suspect, turn out to be 'retro-dictions' in most cases and are poor predictors of future outcomes. Most importantly, cultural attitudes such as mistrust are typically results, and not causes, of the prevalent societal institutions. When political institutions, for instance, do not reliably guarantee property rights, is it any wonder that individuals develop mistrust of strangers or reluctance to take out a loan to develop their land? When a small elite of landowners reap the profits off your labor, is it any wonder that you would not be motivated to work extremely hard?

Instead of these explanatory attempts, the authors make a compelling case based on detailed examinations of wealth distribution throughout history and geographic location that the answer lies in institutions: everything really comes down to the type of institutions predominant in a given society. The prevalent political and economic structures are what matter. If political organization is highly hierarchical, with power concentrated in the hands of a small elite, the economic system will also be highly exploitative, or extractive, as the authors call it. This dual combination of a political system with narrowly concentrated power at the top and an extractive economic system precludes the emergence of wealth whenever it is present. It leads to vicious feedback cycles that destroy national economies and wealth for the vast majority of the population. Conversely, according to the theory, at all times and in all places where political power was and is more broadly distributed and economic systems developed that were more inclusive, wealthy societies emerged.

The primary mechanism for this principle of cause and effect is again feedback, here primarily in the form of incentive. In hierarchical and extractive systems, the incentive structure interestingly is dually unwholesome. On the part of an individual outside the narrow elite, there is absolutely no incentive to work hard and innovate. The social structure makes sure that all the rewards get transferred to the top anyway. On the part of the elite, there is often no incentive to adopt new technologies that would make production more efficient. These innovations often have a way of threatening the very power structure from which the elite benefits. And since the means of production are monopolies, there are no free agents to worry about that might gain an edge by the adoption of technology. The status quo is, in fact, preserved by rejecting innovation.

Again, there is this recurring notion of *self-consistency* in feedback shining through. It is essentially the same idea we encountered earlier in condensed-matter physics. Here, in this context, one might also call it a self-fulfilling prophecy. Individuals act in ways that are consistent with the larger structures and their resultant feedback on those same individuals. Round and round we go, but nowhere is there a contradiction. Individuals behave rationally given the incentives established for them by the social structures set up by the larger economic and political systems. But in acting that way they also support, strengthen and perpetuate those same systems that created the incentives in the first place. It is not difficult to see how once set in place, this pattern is extremely hard to disrupt, precisely due to the self-consistency that keeps it going. The stability of such systems, and the durability of the phenomenon of poverty generally, is thus understood as a consequence of self-consistency: the behavior of individuals that make up social structures is consistent with the feedback on individuals by those same structures.

Contrast this situation with life in participatory democracies and inclusive economies. What are the feedback loops operating here? Simply put, the incentive for individuals is to perform well at work, to educate themselves, to invest in skills and training, to learn to be innovative and productive, and so on. Why? Because companies and employers incentivize those exact attributes by offering promotions, pay raises, and the like to those who do these things. Why do companies gain when their employees value things like productivity and innovation? Why do they incentivize that behavior? The answer is obvious; they have an incentive to do so as well. They gain a competitive advantage over rival companies. Workers intuitively get that concept. At our places of work, we have all had thoughts along the lines of: 'I want to get that pay raise next month, so I will do things that the firm seems to value at the moment. The firm seems to value it because they have identified this as a key component of their strategic plan.' In that way, there is a strong motivation on all levels to innovate, adapt, and adopt new technologies. Only monopolies would have nothing to gain and sometimes much to lose from innovation. There is that self-consistency between levels once more, but this time it resulted from and embodies a virtuous cycle of positive reinforcement.

7.3 Social Beliefs and Downward Causation

Graphite and diamond are both made up entirely of carbon atoms. For a condensed-matter physicist, it seems natural to think that nations are rich or poor primarily because of the structure of economic and political institutions, and not due to any inherent attributes of their citizens. As a physicist we might even be skeptical of the term 'inherent attribute' altogether, since we appreciate that most attributes are 'renormalized' by interactions and feedback. The interactions in strongly-coupled systems usually change the attributes to the point where it becomes hard to say what the isolated constituents would be like in the hypothetical absence of such interactions.

At what age do we become aware of these larger societal feedback loops? For instance, when do we start noticing that we belong to particular social groups, and when do we start making that association part of our identity? These are interesting questions sociologists have been trying to sort out. The short answer seems to be: "Sooner than you think!"

A number of fascinating studies were conducted in Northern Ireland—a region with a fairly recent history of violent conflict. The social fault lines are essentially sectarian, running straight through the Catholic and Protestant populations that make up Northern Ireland. These groups have had different political allegiances and tended to live in segregated neighborhoods. Quite typically, the divisions tended to be reinforced by socio-economic status. Add to it religion, and you get a brew that includes almost every volatile ingredient in the mix: religion, politics, economics, segregation. So when would kids first get an inkling of these fault lines characterizing the world into which they had been born.

It turned out that it didn't take too long for their innocence to start receding. Beginning at age three, Catholic and Protestant kids exhibited "small differences in their preferences for particular people's names, flags and in terms of their attitudes towards Orange marches and the police. Five and six year olds also showed differences in terms of their preferences for particular combinations of colours and football shirts," found the study's principal researcher, Paul Connolly in 2002 [84]. Is that an indication that these kids were aware of the reasons they preferred one symbol over another, or one set of colors over another? According to the same study, a surprising 51% of 3 year olds were able to attach at least some cultural significance to their preference, and a staggering 90% of 6 year olds could do so. Apparently, it doesn't take long for society to pass along prevailing attitudes to its youngest members and perpetuate them in the next generation. The one thing that took a little longer was outright identification in terms of belonging to a particular social group; this typically didn't occur until around age 6 and up, according to the same study.

Northern Ireland is, of course, no exception, and similar findings are reported from across the globe. In fact, naively speaking, if kids as young as three can begin to pick out fairly subtle identity markers such as emblems and flags, we might expect them to definitely discern racial affiliations at that age. So, in a multiracial society like the United States, when do kids first notice race? Fairly recent studies provide a startling answer, one that takes most of us by surprise. The answer seems to be: as early as 6 months.

'What? How does one even test babies that young,' one might be allowed to wonder. Well, it has been known for a while that babies stare at things longer when something seems out of the ordinary to them; when something happens that the baby didn't expect or isn't yet familiar with. So, for instance, a spinning top or a yo-yo and their non-intuitive motion might hold a baby's attention for a good while, which probably explains their continued popularity as toys. And it turns out that babies took longer peaks at pictures of faces that belonged to a different racial group than their own—a finding made famous by Newsweek magazine's provocative front-page header of "See Baby Discriminate" [85]. So babies seemed to be at least aware of the difference. But before we jump to conclusions, we must add that noticing a racial difference didn't yet translate into anything more for the babies. It wasn't yet laden with any particular meaning. By age three, kids were beginning to place additional meaning to that difference, the same study found.

And that "meaning" about social groups usually starts to include stereotypes. An influential study in the 1990s by psychologists Claude Steele at Stanford University and Joshua Aronson at the University of Texas tried to get at the effects that stereotypes have on people in a clever way. In this particular study, they wanted to quantify the effect of negative stereotypes on African Americans, and they selected a group of black students at Stanford for that purpose. Again you might wonder: How can you test whether stereotypes made a difference or not? Don't stereotypes simply exist or not exist? How could you control for this? It's not like you can turn off the presence of stereotypes pervasive in our society. That would be like turning off gravity.

Well, in a sense you can turn on and off stereotypes in the minds of individuals by giving them cues. And that's just what the Stanford study did [86]. The basic idea was to administer verbal tests based on the Graduate Record Examination (GRE) typically taken in the senior year of college; so the test included challenging questions. One group of African American students was told that the exam would be a test for aptitude; this was the "diagnostic group." The test was billed as a "genuine test of your verbal abilities and limitations so that we might better understand the factors involved in both." They were told to try hard so as to "help us in our analysis of your verbal ability." These instructions were designed to trigger an awareness of stereotypes about African American performance in this area.

The control group heard nothing of the sort. They were given the exact same test but were told prior to taking it that the purpose was to understand "the psychological factors involved in solving verbal problems." They were also asked to try hard to "help us in our analysis of the problem solving process." Here the instructions did not suggest that what was being tested was anyone's inherent ability or aptitude, but served some more general, vague, or impersonal objective of studying the processes involved in learning.

The results were nothing short of eye-opening. The control group did significantly better on the exact same test. And the difference wasn't just in the margins, but statistically robust. In one run, African Americans in the control group solved twice as many questions correctly as in the diagnostic group. Just the way the exam was framed, the precise language that was used to describe the purpose of the exam, made a large and statistically significant difference. In fact, the study went further and performed the same test on white Stanford undergraduates, using the same diagnostic and control group instructions. For the white students there was a slight trend in reverse apparent in their performance— the effect either went away completely for them, or had the opposite sign. The end result was that under the diagnostic condition, where the exam was billed as testing ability, African American students performed substantially worse than their white peers. But in the control group, that achievement gap (normalized by previous SAT scores) had vanished entirely (Fig. 7.3).

The authors coined the phrase *stereotype threat*. It refers to the measurable effect that the mere awareness of the existence of a stereotype can depress people's performance on tasks related to that stereotype. The operative word here is 'mere awareness.' It's enough just to be aware that others might hold a negative stereotype about your group to makes a difference. There is no need to have internalized these stereotypes or to believe them yourself. In fact, the study cited above was also able to show that African-Americans in the diagnostic group

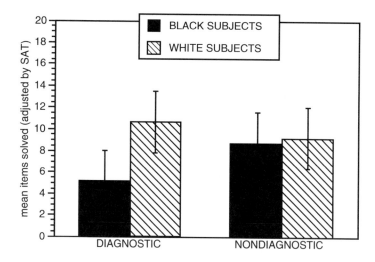

Fig. 7.3 Performance on verbal test by race, with and without the activation of stereotype threat, taken from [86]. The same experiment repeated more recently showed even more dramatic swings

were not only more aware of stereotypes against them, but importantly made additional mental effort to distance themselves from those same negative stereotypes (for instance, by trying to refute them, consciously or unconsciously).

Since then, many other studies have confirmed the stereotype threat effect, with the upshot that it is pretty universal. It always appears in any social group against which there exist societal stereotypes. A series of the New York Times articles in 2013 examined the underlying reasons for a significant under-representation of women philosophers in academia, and again social and cultural feedback figures prominently in the discussion [87].

Now, in science, to really demonstrate a cause-effect relationship between two phenomena A and B, it is not sufficient to simply show that A and B always come together, or that A always precedes B. Some correlation in appearance is not enough; we also look for plausible mechanisms that could explain the pairing of A and B. And that is just what psychologists have tried to do. Yes, the stereotype threat phenomenon is robust in the sense that it shows up in all kinds of situations. But how does it actually work? What makes people perform worse on certain cognitive or athletic tasks when confronted with a relevant stereotype? In other words, what are the downward-causation mechanisms that facilitate this top-down feedback and allow it to function?

Let me just mention a few of the various psychological mechanisms proposed. Some have to do with attention. Having to worry about how others might perceive one's performance takes away cognitive resources that could otherwise by used to solve the task. In situations where the task requires our full attention, as on a challenging verbal exam, this distraction could clearly be detrimental. Any worry about confirming someone else's stereotypes subtracts from our concentration, any threat to our self-image requires a response that again takes our mind away from the task. It is never fun to carry the extra mental or emotional burden of having to overcome "stereotype-activated self-doubt" [86] or to prove someone else wrong about their possible negative stereo-types about you. So clear candidates for mechanisms abound: reduced concentration, shortened attention span, increased anxiety, etc. These are things that are known to reduce performance on tests for everyone, independent of the issue of stereotype.

Another strategy for dealing with stereotype threat, of course, is to stop caring, to "withdraw effort", and to lower one's self-expectation. In the study above with Stanford undergraduates this mechanism probably didn't play a big factor (as the authors argue as well). Students tried hard and didn't lower their expectations of their own performance. In fact, they wanted to do well on the test. The study indicated that they re-read the instructions more often than their white peers. But this mechanism does come into play in high-school settings, for instance, where quite frequently a gradual disengagement with the activity related to the negative stereotype is observed by educators. If a stereotype has the effect of undermining one's self-image, we can simply stop using our performance related to that stereotype as a contributing factor to our self-image. Simply put, we stop caring. Of course, this response has the consequence of most dramatically amplifying the feedback loop.

When we stop trying to excel, we won't excel, and so the stereotype is further strengthened. It is the epitome of a self-fulfilling prophecy. Certain things in our society are the way they are for a reason, but that reason often has little to do with inherent ability on the part of individuals.

A more recent study by researchers at Harvard and the University of Michigan is equally illuminating. Here the authors wanted to see how early in life these mechanisms become effective. In addition, they selected a social group where contradictory stereotypes existed. This ambiguity allowed for the possibility of positive or negative feedback. The study focused on Asian girls and examined their performance on mathematics-related tasks. The idea was that this group of young participants was potentially subject to two conflicting stereotypes. Asians in the U.S. are generally associated with doing well in math, and girls are still sometimes stereotyped as weaker in math. So how was this study going to isolate these different feedbacks?

Well, that's the fascinating and somewhat surreal part, and it is laid out in great detail in the procedure section of the paper [88]. So imagine, if you will, a group of Asian girls of ages between 5 and 7 being led into a room, sitting down at desks, given some Crayons, and asked to color in a picture of a girl holding a doll. The ages of the young participants put them in kindergarten through second grade. Simultaneously, a very similar group of Asian girls of the same age range marches into a separate but identical room, again given Crayons and asked to color in a picture of "Asian children eating with chopsticks out of rice bowls." Finally, there is also a control group, a third group of Asian girls who are asked to color in a picture of a landscape scene. You probably see where this is going: after completion of this preliminary coloring task, all three groups were given identical math tests at their grade level in their separate rooms.

Can you guess the outcome? Yes, the preliminary coloring activity served as a kind of mental cue, a kind of switch, that activated a certain stereotype awareness. And indeed, within the 'gender-activated' group, the girls did significantly worse than in the 'Asian ethnicity-activated' group. The control group was somewhere in between. What this study thus indicated was that kids as young as 5 are already aware of social biases that exist out there pertaining to their group. In fact, it had been known that kids as young as two could reliably discriminate gender in people, skin color as early as 6 months (as mentioned earlier), and ethnicity by the ages between 3 and 5. But the novelty of this study was to show that things went quite a bit further very soon thereafter, that stereotypes started to affect kids' behavior as early as in kindergarten.

Next, the researchers probed whether these performance gaps persisted in upper elementary school and then again in middle school. So they essentially duplicated the experiment with these other age groups, with the only major difference being the way in which the priming was achieved. After all, you won't get very far asking middle school kids in grades 6 through 8 to color with Crayons. So, instead, they gave them preliminary questionnaires to fill out. The questions that appeared inquired about whether the girls had non-white classmates, what race most of their friends were, and so on, or whether their friends were mostly girls or boys, whether any of their friends were dating, and so on. In the control group, the questions were neutral, asking about their likes of animals, and so forth.

In this age group of upper elementary school, something unexpected happened. The trend reversed, and the gender-activated group outperformed all others. But when the experiment was repeated with middle-school kids, the previous trend firmly re-established itself. Negative gender stereotypes again substantially depressed performance on the math test. The conclusion was that the evolution of the feedback mechanisms involved in stereotype threat does not always proceed linearly. In young children, it doesn't appear and then simply increase in strength steadily with time. But by early middle school, it seems that certain patterns have taken permanent hold and stay with people for the rest of their lives. Even by college, this threat is alive and kicking, as we saw earlier.

One important aspect does change qualitatively with age, however. As we get older, we can distance ourselves intellectually from stereotypes, rejecting their validity while acknowledging their persistent power on people's lives. We gain the mental flexibility that young children lack to simultaneously notice the presence of stereotypes and their negative consequences, while at the same time not buying into them. But even as we dismiss them intellectually as unfair generalizations about people, it doesn't mean that we are entirely free from their adverse influence. We're typically still affected by the feedback. As members of society, it is very difficult to loosen that downward control that society imposes. That's not to say it is impossible—there are many examples of highly successful black female mathematicians, for instance, and it would be instructive to study their individual strategies. But for a phase transition to occur, developments can't stay confined to the individual level; it takes a virtuous feedback cycle to take root operating between the societal and individual levels.

How large-scale change gets initiated and unfolds in social systems is a topic of much ongoing research. What condition are necessary to make a system susceptible to global change? It's a pressing question whose answer has potentially far-reaching implications in many contexts. For instance, if we could predict when and how fads get started, which ideas will spread, or what product will become the next bestseller, we stand to make a lot of money. Or think of public health—if we could only predict the kinds of structural factors involved in turning isolated cases of the flu into an epidemic, interventions by the CDC or the WHO could be that much more targeted.

A fairly new and promising approach to such questions involves bridging the gap between mathematics and sociology [75]. Mathematicians have long explored the properties of graphs (also known as networks), graph theory being a vibrant branch of mathematics to this day. A graph is basically a bunch of nodes connected in some fashion by links, and they can be classified according to the rules by which the links are formed. So graphs can be classified as "random," "small-world," or "scale-free," and defined in terms of quantifiable network properties such as "mean path length" or "degree distribution" or "clustering." Sociologists have, of course, accumulated extensive knowledge of the real quality of social networks and social dynamics. As Duncan Watts points out [75], it is not enough to treat social networks as abstract mathematical graphs. The real world is more complicated and models have to reflect that complexity and nuance in order to capture enough of the essential features. They have to be fine-tuned and tailored to the particular social context.

A couple of things complicate matters when modeling social systems. One is that we face two types of dynamics. On the one hand, there is the dynamics on the network, as in the case when information is shared between friends. The other type is dynamics of the network, which involves the formation of new friendships or the severance of old ones. Furthermore, these two types of dynamics are, of course, inherently coupled. Another complicating factor is that there is not one underlying social network in operation at any given time, but that each type of interaction or activity may be characterized by different layers stacked on top of one another. Your network of friends and family may be quite separate from your network of colleagues and work contacts. The upshot is that the details matter; in most instances they cannot be simplified or generalized away.

As Thomas Kuhn reminded us, scientific revolutions most often originate at the boundaries of two established disciplines; they don't typically emanate from their cores [89]. And here too, rapid progress is being made at the various interfaces of graph theory, dynamical systems, complexity theory, computer science, and sociology. The overarching theme in much of this work is not all that different, in its essence, from the lessons encountered in condensed-matter physics: Structure matters! Global phenomena at the system level arise in dependence on network structure. Function often follows structure. How robust a system is against random perturbations often depends on its degree and type of connectivity. How fast a given virus can spread depends crucially on elements of social structure. The efficiency of communication networks relies on the presence of enough nodes of high degree. The ease of finding a new job may be influenced by the number of "weak links," and the rapid dissemination of ideas on the presence of "random" connections. Notions of emergence seem to abound in all directions.

Of course, most of our everyday experience is local and direct. It is a distorted appearance, however, because we do not directly see the social feedback loops at work, nor do we apprehend the social web we are part of. These networks and feedback loops are hidden from our direct sight, and they can only be ascertained indirectly, and yet they guide us in incalculable ways like an invisible hand. We are best at perceiving phenomena that play out within our level of description—at the level of our local network and our individual interactions. We are not as good at picking out processes that happen at 'lower' or 'higher' levels of organization—the level of brain chemistry and neuronal activity, on the one hand, and the level of social systems on the other. Even though a phenomenon may be best described at a particular level, it still needs to be consistent with supporting processes occurring concomitantly within neighboring levels. The layers of complexity are not mutually separate. They can perhaps be likened to a laminated board [77]. There are many layers of description, they do not merge into one another, but instead they are can be visualized as glued together to form a common laminated board.

I find this metaphor of the laminated board quite useful. It charts a middle ground between the extremes of reductionism and strong emergentism. Reductionism tends to deny the validity of the layers altogether; in this view there is really only one layer—the bottom one—into which all the others merge. Strong emergentism, on the other hand, sees the layers as completely separate. Higher-level phenomena are *reified*, as if they existed independently of their constituent parts. The correct view is somewhere in the middle. Phenomena are properly understood at a commensurate layer of description, a description given by an effective theory. At the same time, phenomena depend on, or are correlated with, processes playing out at the level of their parts. The parts have to behave in certain ways in order for the larger phenomenon to arise. The various layers of description are glued together.

Perhaps not surprisingly, totalitarian ideologies veer towards the extremes of social reductionism or emergentism. Communism exhibited a tendency towards exaggerated social realism, via a kind of reification of large-scale social phenomena. Marxist philosophy saw individuals mainly through the prism of stratified social class and economic status. In *Das Kapital*, Karl Marx famously treated individuals as "personifications of economic categories." Not surprisingly, Communism shifted the focus radically away from people's unique dispositions, aspirations, volition and character, and towards society, and its modes of production, its distribution of commodities and capital. It became a cold, highly depersonalized system of thought.

But in contrast to communism, fascist ideology strangely embraced both extremes simultaneously. Germany's National Socialists subscribed to an extreme form of emergentism whereby the nation became a singular monolithic entity to be cherished above all else. Whole nations and 'races' of people were seen as the only real protagonists, and the dynamics between them the only locus containing any real meaning or significance. People were entirely dispensable, their value infinitesimal when compared to the interest of the nation-state. On the other hand, upon further examination we can also discern an adherence to extreme social reductionism. Whole nations and classes of people were considered inferior because the individuals belonging to those nations or classes were inferior. In other words, the source of the perceived inferiority lay in the individual traits shared by all members of "non-Aryan races." And in another leap of biological reductionism, these traits were believed to be consequences of the differences in genetic makeup.

In fascist ideology, it's not as if the "fault" was confined to the level of nations and their structural organization. If that had been the case, then members of ethnic minorities who had lived in Germany and integrated into German society would have been welcomed. That would be the logical conclusion if one was concerned only with characteristics found at the large-scale plateau of nations and societies. But, of course, the opposite attitude actually prevailed.

Take the many German Jews who had fought valiantly in the German army during World War I, and were in all things German patriots. They had studied at German universities, and through hard work and talent had entered respected professions in Germany. The famous chemist, Fritz Haber, considered himself German first and Jewish second. For a long time, he was willing to de-emphasize his Jewish heritage to assimilate ever more assiduously into traditional German society, which the increasingly intolerant climate of the early 1930s demanded. Soon after the Nazis took over in 1933, all of that became irrelevant. Jews were banned from most professions, had to resign from professorships, judgeships, from the civil service, couldn't marry Germans, couldn't even employ Germans. In that year, they were basically forced to retreat from public life altogether. The subsequent years witnessed ever greater crimes against humanity. This was the work of a mentality that subscribed fanatically to social and genetic reductionism coupled with extreme racism.

History has taught us over and over again that we have to resist treating collective social phenomena, such as the nation-states, as if they were solid entities existing in themselves. They do not exist independently of the individuals that participate in them. Rather, they come into existence as a result of the collective imagination of many individual people. They are nothing other than a collective mental construct, a collective dream if you will, that resides in the minds of the many individuals participating in it. Nowhere else does 'the nation' exist. It does not have a separate reality independent of the shared imagination of the many. A nation cannot be touched by our hands or seen with our eyes—it is an idea at its core.

However, as we saw earlier, we must also resist the opposite conclusion to which we might now be propelled, namely that collective social phenomena do not really exist at all. We must find a middle ground. Only because things like institutions, corporations, or nations don't exist as solid entities does not mean that they can't function in very real ways. All phenomena, whether they be social or physical, are essentially emergent in nature. They function at their given scale and level of description in reliance upon mechanisms for downward control onto their substrate (at the level of parts). From one point of view, the whole is nothing but a collection of its parts, but from a complementary point of view the parts behave in certain ways only due to the emergence of the whole. An acknowledgment of this delicate feedback cycle between the levels of description lies at the heart of emergence.

Chapter 8
Larger Lessons

8.1 Science's Organizing Principles

A reductive view of science usually leads to categorizing the various sciences in terms of a strict hierarchy of significance. At the top are the social sciences, then comes psychology, below it follow neuroscience and biology, then chemistry, and finally physics at the bottom. In this scheme, it must be quickly added, being lower on the totem pole is considered better, more fundamental, closer to the final theory of everything. Conversely, a higher position suggests adjectives such as derivative and applied. Within this logic, psychology is just applied biology (biology of the brain), biology is applied chemistry (the chemistry of cells), chemistry is merely applied physics (the physics of molecules). Even within traditional science disciplines, a hierarchy is set up. For instance, within biology there is microbiology at the bottom and ecology, botany, or physiology near the top. In physics, elementary particle physics is near the bottom and atomic, molecular, condensed-matter, astro- and bio-physics near the top (roughly in that order). Reductionism forces the sciences into strict evaluative schemes and compartments.

It is not hard to see that this view creates tension between the sciences. It breeds unwarranted arrogance on the part of scientists in fields near the bottom of the list. Here it encourages unrealistic attitudes

© Springer International Publishing AG 2017
L.Q. English, *There Is No Theory of Everything*,
DOI 10.1007/978-3-319-59150-6_8

in them by suggesting that their expertise must also extend into the fields of their colleagues further up the chain. Conversely, it engenders hostility among those scientists, who tend to cultivate an attitude of dismissiveness towards disciplines at the lower levels, caricaturing them as toying with simple systems that could never capture the complexities encountered in their work. It certainly does not foster a spirit of collaboration and cross-fertilization of ideas across boundaries.

Emergence opposes that hierarchical view and in fact dispenses with it altogether. There isn't really an up or down. There is only scientific description of processes and the discovery of law-like regularities in systems differentiated by scale, complexity, or organization. Meaning is found in the effective theories or models developed to explain a particular phenomenon at its appropriate level of description. The sciences are not separate but make contact with one another on lines of continuity. A phenomenon at a larger scale is best described at that scale, but it is supported by processes at shorter lengths. These processes are typically not in a one-to-one relationship with the emergent phenomenon and often can't be used to derive it, but they participate in its appearance nonetheless.

It is clear that this view is not only closer to reality, it is also encourages cross-disciplinary interaction. It allows scientists with different backgrounds and areas of specialization to come together to study a particular phenomenon from different perspectives, without notions of superiority or inferiority of approach or interpretation getting in the way. It is a supremely pragmatic view.

As an example, take meteorology—the study of weather and weather patterns. At what levels can you approach weather? Well, classical physics enters into the picture via fluid dynamics and via effective forces experienced by parcels of air in motion, such as the Coriolis force and pressure gradients. But much more needs to go into these models, a lot of it in the form of heuristic descriptions of processes from atmospheric science, such as cloud formation and precipitation. Computer science enters in, as the resultant equations can't be solved analytically, but need to be simulated numerically. Here, a whole set of related but self-contained issues arise in terms of numerical strategies and recipes. How best to discretize the earth's atmospheric volume, for instance, so as to transform the governing differential equations into a form a computer can handle? What algorithms work best to evolve the system forward in time? How best to farm out the task to many computers and gather the output effectively?

But computer models also need tons of input. So atmospheric science is crucial again in supplying data that capture the large-scale conditions of the atmosphere, like pressure, temperature, humidity at various weather stations. Oceanography enters in, as the air picks up moisture from the oceans and exchanges heat energy with them. Geography becomes important, as the topology of the land forces air currents in certain directions where they cool or warm, lose or gain moisture, and so on. When tornadoes or hurricanes form, an empirical knowledge of how they pick up, as well as dissipate, energy in dependence on other atmospheric conditions is obviously crucial. An empirical understanding on what factors might force a hurricane to turn inland is clearly relevant. The list goes on and on.

What's fascinating is that contemporary science is full of these kinds of examples. Astronomy, as a second example, draws on an equally rich and diverse array of scientific branches. It makes meaningful contact with geology when studying planets, their surface morphology, their tectonics and interior composition. Think of the rover that NASA sent to Mars and all the sensors with which it was outfitted. Many of these sensors were designed to pick up traces of chemical compounds, so chemistry now takes a leading role. Accurately describing stellar births and deaths within galaxies involves intricate knowledge of thermodynamics, nuclear physics, and quantum statistics, for instance. Galaxy centers usually feature massive black holes that distort the space-time around it according to the laws of General Relativity, and that create enormous temperatures in an area around the galactic center by accelerating matter towards it. Now atomic and molecular physics enters the stage contributing its detailed knowledge of spectral lines and their dependence on temperatures and pressures. Furthermore, spectral

lines can indicate the presence or absence of electric and magnetic
fields, and so electromagnetic theory (in conjunction with quantum
mechanics) comes into play. Sometimes intense magnetic fields are
produced which then interact with the surrounding plasma, and this
interplay creates new large-scale phenomena such as particle jets.

Such examples of the ways in which the physical and life sciences nowadays have to interact to solve complex problems give us
another facet of emergence—one which plays out on a meta-level as
an evolution of science practice and organization. The point is that
reductionism is really an inadequate framework for understanding how
much of science is actually conducted these days. These endeavors
are not characterized by strict hierarchies but by a collaborative spirit
that approaches a given phenomenon from multiple, diverse, sometimes
disparate, yet often complementary, perspectives. The fullest meaning
is obtained via the accumulation of the separate contributions that each
perspective yields. Science, as it is practiced presently, appears less like
a ladder and more like a tapestry. Or, to employ a previous metaphor, it
is like a laminated board.

8.2   Science and Religion

We can broaden the discussion to science's proper relationship with
other areas of human inquiry. First, where are the boundaries? Science
is understood to be about things that can be registered by our senses,
or measured by the extension of our senses—scientific instruments.
Brain activity can be measured using electrodes or MRI, but states of
mind cannot, as they are by their very nature subjective experiences.
Therefore, we have hit the dividing line, the boundary of science.

We might think that this dividing line cedes vast territory to science
and leaves only a small area to everything else. When we think about
it for a moment, however, we realize that for us virtually everything is
the stuff of mind. Granted, through our senses we make contact with
the outside world, but even this sensory input is filtered by the mind and

converted into an inner experience. This leaves ample room for fields of inquiry that do not start from the material and the physical. There is plenty of room for understanding the world we live in through literature, art, religion, and so forth.

The relationship between science and religion is perhaps the most salient, because of their tangled history and the ongoing public debates surrounding it, but also because both make claims that seem to encroach on the other's domain. I often see the bumper-sticker "Coexist" taped to the back of cars where each letter is in the form of a symbol representing a particular faith. The question on many people's mind is: "Can science and religion peacefully coexist?" Emergence shows us the way by which that can happen.

How do they encroach on each other? Well, science has expanded dramatically over just the last 50 years. To many people within science and outside of it, there is the implicit assumption that ultimately every aspect of our reality will be explained through its lens. These days, many people profess that everything ultimately has a physical explanation and that there are no phenomena that cannot be measured by instruments or seen by our senses. In fact, if this last proposition were to be taken seriously, then by default everything is science. But is that the correct starting proposition? Why should we, in the words of theologian van den Haag,

> make 'truth' equivalent to 'scientifically demonstrable at least in principle'? Do the scientists actually have the cosmos by the tail, or do they define the cosmos as that which they do have by the tail?" [90]

Of course, even the staunchest proponents of science would concede that there are things that cannot be measured by instruments. Feelings, thoughts, perceptions, dreams, faith—none of them can be directly measured. But that is usually where the concession ends. Because immediately thereafter they would add that these things can be indirectly measured, at least in principle. They would assert that feeling states have a material basis in brain activity which in turn can be picked up by instruments. Thus, their conclusion is swift: everything falls under the purview of science; everything reduces to a physical description.

But can a feeling of empathy, say, really be reduced to a brain pattern obtained by MRI and displayed as a color map on a computer screen? If we only had that screen pattern (and we hadn't asked the person what she was feeling), would we know what it corresponded to? Maybe, but only because of a correlation that was established at one point by observing the pattern and simultaneously asking the person. Does the pattern give us any independent insight into what compassion is, what it feels like? Not really. Further, would we understand what triggers it? Perhaps it is triggered by other preceding brain activity or patterns. But why? There is no meaning to be found at this level of description. But meaning is very easy to be had at a different level of description: empathy tends to arise in the mind when we see people we love suffer. There—how simple, and yet how beautifully predictive!

It is a false science that claims to be able to capture the true meaning of what is essentially subjective experience. In this arena, science— properly understood—is out of its league. In the words of theologian Paul Tillich, "If representatives of modern physics reduce the whole of reality to the mechanical movement of the smallest particles of matter, denying the really real quality of life and mind, they express a faith, objectively as well as subjectively." [91, p. 82] Stated another way, materialist reductionism with respect to phenomena of mind is not science but a kind of faith itself. It looks and feels more like dogma.

If we take emergence seriously, we have no problem with the state of affairs: science and religion occupy disparate dimensions of human knowledge, even as both play out on society's stage. *Meaning* only ever arises at a certain level of description or analysis commensurate with the phenomenon under question. At the same time, however, the phenomenon is not insulated from other layers of description. No, it causes, initiates, depends on, or co-emerges with certain processes in those layers. Mental cognition does go along with processes in the brain. But its effective description, its essence, is not found there. Rather, it is found within the world of subjective experience, the world of the mind. The mind carves out its own separate reality that does not reduce to physical states, but only correlates to processes in the physical substrate. Again, the emergence metaphor of a laminated board is useful. Only because layers are connected does not mean that they lose their functional reality.

In this way, neuroscientists can continue the fascinating and important work of studying the brain and its function, and religion does not have to feel threatened by that investigation, because ultimately it cannot replace or subsume the spiritual or religious realm. Both have meaning and justification for their existence in their respective domains. Both can acknowledge the other's successes. And both can coexist peacefully without really overlapping. A grand unification, as perhaps envisioned by *natural theology*, seems implausible.

In light of emergence, Paul Tillich's view of science and religion occupying different dimensions of knowledge is particularly compelling. What is envisioned by emergence are separate domains that can coexist without contradiction by virtue of having become emergent *protectorates*, and which reveal emergent truths that don't swap over into other domains. Science and religion can't contradict each other by default—they don't compete in the same arena.

In a more abstract sense, science and religion actually have something aspirational in common. Both postulate the existence of universal laws, and both contend that these laws operate regardless of how someone may feel about them. Astrophysicist Neil deGrasse Tyson recently put it bluntly: "The good thing about science is that it is true whether or not you believe in it." [92] Buddhists speak of the law of Karma in similar terms, as do Jews and Christians with respect to the Commandments of the Old Testament, for instance. Physicist and theologian John Polkinghorne has emphasized in his writing the communality of philosophical outlook at the heart of science and religion, even calling the two "colleagues in the common quest for truth" [93, p. 33], and asserting that both "can lay claim to the achievement of a degree of truthful understanding that warrants their insight being described under the rubric of critical realism." [93, p. 15] Realism, in this context, is meant in the limited sense of referring to the scope of laws as transcending the merely personal, not their essential character.

There are, however, strains of contemporary thought that see all truths as socially constructed and culturally subjective. For many postmodernist thinkers, there is little direct meaning in the discoveries of science (or the tenets of religion, for that matter). Instead, the claim is that such "discoveries" must be understood entirely as constructed, thereby illuminating only the prevailing societal ideas or the intellectual consensus of the times, but nothing that goes positively beyond cultural history. Scientific truths are relegated to an expression of socio-historical opinion, and they certainly do not pertain to "the real world," however conceived. The reason cold fusion was dismissed in the 1990s had nothing to do with the energy scales involved in nuclear fusion, and everything with the formation of a consensus within the scientific community, the thinking goes. More recently, "post truth" science denial has become fashionable on both extremes of the ideological spectrum. What does one make of this critique of science from postmodern circles?

Particle-physicist and Nobel laureate, Stephen Weinberg, considered the persistence of such outlandish views partly a legacy of the same *positivism* that in an earlier age contributed to the erroneous rejection of atoms (by Ernst Mach and others) [94, p. 188]. He may be right in identifying one source of the problem, but emergence offers a simple solution. From an emergence perspective, it is actually fairly easy to counter this postmodernist science skepticism and navigate the "controversy." On the one hand, it is of course true that science is a human endeavor and as such its precise trajectory is affected by "the merely personal," as well as by larger societal trends and opinion. Science does not live in a social vacuum. Nowadays, for instance, the set of topics that get the most attention in science depends on external things such as grant announcements, availability of funding streams, the possibility of practical applications that address a societal need, and so on. Which papers get published depends on the peer-review process, and therefore on the judgments and proclivities of the larger scientific communities. So clearly there are a host of such mundane aspects that influence how science is carried out in practice and how it advances.

But the content of scientific work—the measurable quantities, the models and theories, the explanatory frameworks and unifying principles—transcends this level of description. It's the same theme of emergence that we have encountered time and again. On the one hand, science and society do not occupy entirely separate spheres, but come into contact with one another. On the other hand, this fact alone does not imply that science can be reduced to a social phenomenon. The sociology of science is a legitimate field of study, but it cannot be equated with science itself.

8.3   Religion and Emergence

Stephen Weinberg once lamented that the more he understood about the physical world and its laws, the less *meaning* there appeared to be contained in them [94, Chap. 11]. If we follow reductionism to its logical conclusion, meaning indeed proves elusive. Ultimately, all there can be are the random motions and interactions of a myriad of elementary particles. All other elements of our experience—our thoughts, aspirations, dreams, free will, up to our spiritual insights and religious beliefs, must appear like mirages that dissolve under greater scrutiny, having been exposed as mere make-belief. They are just derivatives of some more fundamental and materialistic description of nature. It is a pessimistic, even nihilistic, view of humanity and of the world.

I have many colleagues in physics as well as in the other sciences who have found ways to harmonize their spiritual or religious beliefs with science. Not surprisingly, I also have colleagues, friends and acquaintances who are sympathetic to scientific materialism. Some of them will argue for this point of view passionately, as if to convince others out of their "self-deluded fiction." I wonder, though, if they have really fully subscribed to and internalized their own paradigm?

If they had, would there be any reason for lively discussion. Even my differing view would be traceable to particular differences in my brain structure. What would be the reason for trying to change it? Why argue for or against anything if neither you nor your discussion partner actually possesses *free will*? It is similarly amusing to hear about a neuroscience study claiming to have proven the non-existence of free will, when it was always clear from the beginning that such a notion would have no room within the materialist and reductionist paradigm. Free will is essentially incompatible with reductionism. For biologist Stuart Kauffman, human agency should therefore be understood as a instantiation of "ontological emergence" [95].

Let's take the notion from neuroscience that all of our thoughts, feeling, mental states, including also our moral sense, are direct products of brain function or neural processes. Perhaps there is a particular region in the brain that decides for us and controls our actions—a kind of decision center where sensory information gets processed. The question then becomes: how are these decisions made? If "we" don't have free will, then does this decision center—this "homunculus" inside our hippocampus—have free will? Or is it deciding in a very mechanistic or algorithmic manner, whereby a set of inputs generates an output. In this scenario there is no room for free will or notions of personal responsibility. On some societal level, we must not really subscribe to this view, otherwise it would not make sense to hold criminals accountable for their actions in a court of law (in the sense we do not for the mentally insane); it would not make sense to be disappointed with our kids when their conduct does not live up to our expectation.

Interestingly, one can make a compelling argument that human agency is actually necessary for science to exist in the first place. The scientific process, after all, says that we formulate a hypothesis about an observed phenomenon, and then we subject the hypothesis to experimental scrutiny. We weigh the evidence and based on it we confirm or reject the initial hypothesis. But notice what assumption lies at the heart of observing a phenomenon, proposing a hypothesis, devising experiments that speak to it, weighing the evidence and finally arriving at a decision as to the validity of our hypothesis. Doesn't it require that we have a choice to make? That we have the creativity to design experiments, that we can freely contemplate the evidence for and against, and that we can sort out the various inferences freely, objectively, and unencumbered by biological dictates? Only free will renders this process of exploration and contemplation meaningful. Robots cannot engage in this kind of process. They could have been programmed to arrive at certain unrealistic conclusions no matter what the evidence says, and they could even be programmed to appear to 'believe' in those conclusions. There is no guarantee whatsoever that either their observations or their conclusions are accurate reflections of how things really are.

So what do we make of a scientific study that claims to have discovered the lack of free will, that claims to have exposed human agency as a fiction? Well, if we took this claim at face value, would it not then undermine the very science behind those claims, and therefore place the claim itself in doubt? The position seems somewhat self-negating. In the words of William Newsome,

> If mental processes are determined wholly by the motion of the atoms in my brain, I have no reason to suppose that my beliefs are true [...] and hence I have no reason for supposing my brain to be composed of atoms. [66]

Again, material reductionism does seem to imply a profound lack of genuine human agency. For obvious reasons, this is a very unsettling proposition for most of us. Furthermore, it is also not reconcilable with spiritual or religious views, for it is precisely this agency that spiritual and religious traditions fundamentally rely upon. They are built on the basic premise that through our intention and choice, we have the power to 'better ourselves' and transform our lives. Of course, that doesn't refute reductionism—only because we don't like the implications of a particular world view, doesn't yet mean it is incorrect. But we can now appreciate that emergence holds the potential of salvaging notions of free will by shielding them from reductive deconstruction. It is an indispensable idea.

Reductive interpretations of religion as a phenomenon have become fairly popular, at least in Western thinking. One such strategy is to relegate religion to the purely psychological. Religion is seen as nothing other than a reflection of our psychological makeup or a means to fulfilling a deep-seeded psychological need. What do we make of this approach from the perspective of emergence? We freely grant that religious experience has a psychological substrate. But despite their psychological correlates, religious experiences do seem to be best comprehended and described in the language and symbols of religion. In that sense, we may be skeptical that psychological analysis alone can capture the essence of religious faith and experience. Just as human psychology is based in human biology but transcends it, so it would seem that religious experience cannot be reduced fully to psychology. A tell-tale sign would be if spiritual or religious practice were able to modify human psychology. Via such *downward* action, it could be argued, religious practice could carve out an autonomous experiential reality from which subsequent novel phenomena then naturally arise, and these would then be meaningfully understood only in reference to that practice.

Another reductive approach operates at a societal level and sees religion as a purely social phenomenon. Again, there is no denying that religious practice can have a profoundly social character and that it can shape social patterns and culture. Similarly, it is granted that religious institutions participate and are subject to lager-scale social dynamics. Religion is decidedly affected by society, and societies have been profoundly shaped by religions throughout human history. The existence of these connections and influences, however, does not make one a subset of the other. This is a recurring lesson of emergence. If we think of religion as occupying a certain layer of reality, then we can acknowledge that developments within that layer of description are supported by concomitant processes in a neighboring layer, without it losing any of its own functional reality.

To start our somewhat informal exploration of emergence and religion, let's ask ourselves what we would do if we wanted to learn about one of the major world religions unfamiliar to us? We could take the college-course approach, one that is rooted in recorded history and archeology with its focus on textual sources, artifacts, historical figures, iconography, and so on. No doubt, this line of investigation would be very fruitful, and you would eventually emerge with a deeper understanding of the socio-historical factors involved in the appearance and evolution of the religious tradition in question.

The historical origins, the sociological circumstances, the prevalent political structures and cultural context within which important religious figures operated—those things are clearly germane. They might be a good place to start. Nonetheless, one could argue that they are ultimately secondary to an experienced reality of the faith. The fact that religious faith goes beyond the historical, and indeed beyond the conceptual, is often cited by practitioners as a key source of its power and transformative impact for them.

Furthermore, it seems similarly plausible that if one engaged religion exclusively on the level of doctrine, on the level of commandments, quotations, elements of cosmology or prophesies, one would again miss the mark. They are important defining features, but by themselves they still do not yield direct insight into religious practice and experience. Religious practitioners almost invariably think of their religion as more than just another philosophical system to be understood intellectually. As Paul Tillich put it, "faith is the act in which reason ecstatically reaches beyond itself." [91]

The phrase *living faith* is frequently used in this context. But how could we discover elements of the living faith, when such faith is ultimately rooted in personal experience? Short of becoming a practitioner oneself, as a kind of proxy we could try to get a sense of the ways in which exercising the religion transformed the lives of its practitioners, by their own account and that of others around them. One could search for common features within the collection of such individual experiences.

A concrete example may be helpful at this juncture, and Harriet Tubman's remarkable life story may serve as an apt illustration here.

To briefly summarize, after escaping slavery herself in 1849, Harriet decided to cross the Mason-Dixon line into the South nineteen more times over the course of the next 11 years, at great danger to her own life, to help free hundreds of slaves and guide them to safety. Harriet was an extremely religious woman, and her faith was heightened by lucid day-dreams. Triggered by an old head injury inflicted on her in an act of violence, these frequent trances took on a spiritual meaning. She described them as visions from God, and as a means for her to communicate with God directly. They became of deep mystical and spiritual importance, but she also credited them with practical decisions about such matters as which routes to take [96, 97].

In one sense, it is not hard to see that Harriet Tubman needed to believe in divine guidance and providence. If she hadn't, it is unlikely that she would have attempted, or succeeded in, her improbable rescue missions. Harriet certainly credited her faith as a necessary condition for her success as a conductor along the underground railroad. Even before that, her Christian faith likely allowed her to transcend the dehumanizing view of her by others, and to see herself and others as endowed with divine potential. In this way her faith created a transcendent reality and had what we might call functional causal power. Thomas Garrett, a prominent figure in the underground railroad and trusted financial and logistical supporter of Tubman's, recalled: "I never met with any person of any color who had more confidence in the voice of God, as spoken direct to her soul." [96]

A skeptic might ask: "Was there really a God out there that spoke to Harriet?" But would anyone really expect to find a deity made up of matter, of atoms and molecules, or of light particles? Would we really entertain some physical manifestation of sort that we could interact with through our ordinary senses of sight and touch? Probably not! Some physical force that can directly re-route neural activity in our brain to induce the perception of an image or sound? Unlikely! A non-physical power that we can therefore only apprehend subjectively within the mental space of our experience? Now we are getting closer—most religious people tend to think roughly in those terms. How else could one become aware of, apprehend, or know any element of religious truth other than within our experience. Christian theologian John Hick expressed it thus:

> We can know only with the aid, and by the means, of the cognitive equipment with which we are endowed. We cannot step outside our nature in order to examine it from without; for wherever we step our nature steps with us. In short all our cognition is relative to ourselves, and contains an inescapably subjective element. [98]

One element of the "inescapably subjective element" of our cognition that appears to be indispensable to an apprehension of, or an encounter with, the divine is faith. Some theologians have objected to the view of a mutual dependence linking faith and religious truth (be it an encounter, an apprehension, or a spiritual realization) as being circular. Raziel Abelson pointedly expressed this criticism of circularity thus:

> Those who lack faith cannot hope to encounter the divine objects of faith. This seems to get us into a methodologically vicious circle, such that we cannot reasonably have faith without first having the Encounter, and we cannot have the Encounter without first having faith. [99]

In other words, faith relies on experience, but experience presupposes faith. Is this circularity an indication that something has gone awry with our concept of religious faith? To the contrary—the circularity is arguably an essential feature, and one which allows a novel phenomenon to emerge.

Let's back up and briefly recall an example from science that we encountered before. What happens when a speaker and a microphone are brought into close proximity to one another? A loud and shrill noise is produced via feedback. The noise is an experimental fact to which all of us can testify. But if we had never witnessed this phenomenon, and tried to reason out if it could happen, we might be led astray by the following line of argument.

We might say: "We cannot have a sound produced by the speaker without first having an input at the microphone. And we cannot have an input at the microphone without having a sound produced by the speaker. Therefore, we postulate that nothing will happen, that certainly no shrill noise can manifest." But this conclusion would be in direct contradiction to what is actually observed. Where was the fault in our logic?

What we missed was the interplay of feedback, noise and amplification, as discussed in previous chapters. So perhaps the sound-system feedback example is an apt metaphor for the relationship between faith and religious truth. If someone manages to develop some faith in the absence of any experience of the object of that faith (what we have termed truth), perhaps that faith allows that person to get a small experience of religious truth—an inkling, but not deep experience, not an epiphany or full realization. Nevertheless, this inkling is enough for them to increase their faith, because they now have something to work with. They might be able to use that inkling to extrapolate to the hypothetical possibility of greater insights. Thus, their faith is strengthened. This increased faith in turn allows them to deepen their experience, which again increases their faith. A virtuous cycle would have been initiated with the end result of a direct experience of an element of religious truth (Fig. 8.1).

Starting at least with Augustine of Hippo [100] and later echoed by Anselm of Canterbury and others, Christian theology has advanced and stressed the notion that faith always precedes religious understanding (rooted in one's experience of truth) and is primary in this sense. Others have coined the phase "circle of faith" [101], implying the importance of stepping into a mental space free from doubt that will produce its own results. In Buddhism, faith is often defined as a mind that opposes our tendency of perceiving faults in holy objects [102, pp. 104–106].

The point here is not to imply anything along the lines of 'in the final analysis, religion is simply a psychological feedback loop.' Of course any aspect of our cognition has a psychological basis. But the statement is too reductive of an interpretation. Instead, the claim is that it is extremely difficult, if not impossible, to make meaningful statements about a religion from the outside looking in. One has to step inside the "religious circle," and engage with its teachings, practices, iconography, and so on, to get an authentic taste of it on an experiential level.

Fig. 8.1 The trefoil knot as a visual metaphor for feedback loops. As we have seen repeatedly, feedback also operates between different levels of organization and thus generates novel elements of reality. Faith and knowledge could be thought of as forming a virtuous feedback cycle

The objection could be raised that any system of thought, even harmful ideologies, could be internalized in this manner, and so there is nothing unique to religion in this description. It is, of course, true that ideologies have their own feedback mechanisms—adherence and "faith" in certain ideologies can construct personal realities and bring about manifest results. It is clear that an internalization of fascist ideology will consequently create its own distorted reality through which external situations or events are interpreted.

So it is granted that any system of thought, when thoroughly internalized, can function in this manner to create manifest results and experienced realities. But what kind of results do we seek? One could argue that the realities that emerge through authentic faith in, and correct practice of, any of the world religions are qualitatively very different from, say, political ideologies. A defining aspiration of religious faith is to draw closer to 'the divine', to 'the transcendental', and away from ordinary self-concern and self-conception. Practitioners of almost all religions strive to move away from their normal self-centered way of interacting with others, aiming instead for a more universal perspective. The world religions are enormously diverse in doctrine, practice, and worship, but a case could be made that this much, at least, forms a common denominator.

The idea then is that a new state can gradually or spontaneously emerge through this feedback between faith and experience, one that has pinched itself off in a kind of 'phase transition' from the previous one. If someone has made this religious leap, emergent phenomena arise within this new state. What are those emergent phenomena? One is novel interpretation, appraisal, and appearance of the external world and events within it, and a second is novel patterns of behavior. We can speak of appearances and modes of behavior that only arise in the mental state that religious faith has created, and are inconceivable outside of this state. How could someone like Harriet Tubman, who had endured so many horrors and traumas over her entire life, first find the will to escape into Pennsylvania, and then, having reached safety, decide that she would venture back into Maryland again and again to lead others to freedom? From where does such strength and fearlessness derive? The ordinary rules governing human behavior seemed to have been suspended; they didn't apply to her.

Again in the words of John Hick, "A sufficiently powerful spiritual catalyst may cause a total reapperception, changing a man's entire view of the world. Such conversion, whether gradual or sudden, forms or reforms the personality around a new center." [98]

In the case of scientific theory, we saw how different layers become effectively *shielded* from the encroachment of laws operating at ever smaller scales. It seems natural to extend this principle of shielding to religion as well. Here we would contend that such things as religious experience, spiritual insights or moral values are not reducible to human psychology or neurobiology. But, interestingly, we also encounter interpretive schemes that re-cast religion in terms of phenomena at the arguably larger societal scale. Even so, we can likewise label this general approach reductive. The reason is that again religious practice and the mental phenomena arising from it would be seen as mere epiphenomena—their true origin claimed to lie outside the proper realm of religion. But we have argued that there is no way to capture the essence of religious or spiritual practice from outside of its circle, since by definition this essence is contained inside the experience of the practitioner—only there does it acquire meaning. The words "larger" and "smaller" (referring to physical scale) should be replaced in this context with "inside" and "outside." The tools with which to explore the spiritual or religious worlds are not detectors and instruments on the one hand, or theories of social organization and institutional norms on the other. Instead, the tools are the minds with which we explore our inner life.

Epilogue

Imagine, if you will, going out for a swim in your favorite pond on a warm summer day. As you dip your feet into the pond, the water feels cool and refreshing. Diving in head first, your hair gets wet and dense, your skin starts absorbing moisture, your bathing suit rapidly soaks up the water. Buoyancy makes your body feel lighter in the pond; when you lie on your back, you can even float with minimal effort. This is your experience of the pond, your reality of it. Now think of other creatures with which you share this pond. On the water surface you notice a small fruit-fly struggling but unable to free itself. The surface tension of the water surface has this insect trapped; it utterly fails to extricate itself from the sticky membrane. You now spot some water spiders not too far off. They walk on the water as if it were a trampoline. Their agile feet make tiny dents on the water surface which are then optically magnified on the pond floor by the way light gets refracted from them. Your gaze wanders as you notice some fish gliding near the bottom of the pond. The water completely engulfs their smooth bodies and streams across their gills, which extract the dissolved oxygen. With quick strokes of their tail fin, they slosh the water around and turn their bodies abruptly in sharp angles.

Even though from a physical perspective, we know that the pond is a vast collection of interacting water molecules, different worlds arise from it for the different creatures that inhabit it. The fruit-fly's

© Springer International Publishing AG 2017
L.Q. English, *There Is No Theory of Everything*,
DOI 10.1007/978-3-319-59150-6

experience of the pond differs drastically from that of the fish, and for the water spider the pond is another thing altogether. When we think about it, the various realities of the pond have little in common with one another and do not much overlap. Each experience is like a world onto itself—a seemingly self-contained bubble that has walled itself off from the surrounding medium.

And within each bubble, we can observe distinct phenomena, and discover idiosyncratic rules that govern them. Yet, each bubble also makes contact with the same microscopic layer underneath it (the water molecules), as well as the macroscopic layer above it (the pond). In this slightly psychedelic metaphor, then, the bubbles represent the many possible worlds, each novel yet arising from a common substrate, and each featuring its own emergent truths.

Come to think of it, to some degree all the examples we encountered in this book shared this feature of "separate, yet not separate." Think back to phase transitions characterized by a global reorganization of microscopic constituents leading to novel and unexpected macroscopic behavior. Here we saw how the whole often defines the nature of its parts. Think back to the formation of large-scale convection patterns and their strong downward control on the microscopic substrate, or alternatively to the swarming behavior of fish. Remember back to the surprising success of effective theories that are both loosely motivated by microscopic processes but also ignore the details.

Through these examples we appreciate a recurring theme and a common thread: a layer of description can make contact with neighboring layers without a simultaneous loss of its practical utility or its functional distinctiveness. Chemistry and physics are not completely isolated from one another. They do not live in separate universes described by different fundamental laws, or describing entirely unrelated phenomena. No, the two fields share a long and somewhat blurred boundary. Both fields provide a description of the electron's states within atoms and molecules, for instance. Yet we cannot go as far as to conclude that the two disciplines are really just one field of inquiry. The seemingly endless rules of organic chemistry cannot all be derived from the quantum mechanics of the covalent bond.

The same principle is at work beyond the physical sciences. It certainly extends into the biological sciences, where it is clear that microbiology is a science irreducible to chemistry in the sense that its rules cannot be strictly deduced from chemistry. Yet it is built on the foundation of chemical processes. The cell "appropriates" chemistry for its own purposes. In a striking illustration of downward control, it forces particular bio-chemical processes to occur at certain, well defined times and suppresses others. Despite this downward control, however, it would be a mistake to regard the cell as a singular entity or phenomenon. In the final analysis, the cell is still just a collection of bio-chemical processes and constituents. The point is that this collection has organized itself in such a way that it can act back on and modify the behavior of its parts in a self-consistent manner. And out of this feedback, a functionally separate and new phenomenon emerges that we call 'cell.'

Sociologists have long appreciated notions of emergence and the importance of non-reductive explanations. I have argued that the principle applies even more broadly and informs questions about the limits of science and the relationship between science and religion. Here again, we can appreciate that while religion certainly makes contact with neuroscience and psychology on the one hand, and sociology and history on the other, religious or spiritual meaning is not found within those descriptive layers, but must instead be found within the layer of religious experience. A grand unification between science and religion within a kind of natural theology does not seem plausible, as the two modes of inquiry represent a difference of type—one concerns itself with the objectively measurable and the other with the subjectively experienced.

For me personally, notions of emergence have long accompanied my research endeavors, often without me being fully aware of the fact. In physics many of us have an unfortunate habit of focusing on mathematical formalism, experimental procedure, or narrow physical interpretation, to the exclusion of anything broader. When talking to colleagues about our work, we emphasize those technical aspects and purposefully leave out broader connections. It's almost like an unwritten rule: "thou shalt not discuss or speculate on philosophical implications or ramification of your work." This even happens to famous physicists—David Mermin once related that he was met with some hostility when talking about the meaning of quantum mechanics. So I might be forgiven for having missed the forest for the trees.

In hindsight, however, emergence was never too far away. Aspects of my Ph.D. work on patterns formation in anti-ferromagnetic crystals had this flavor; we showed, for instance, that the macroscopic shape of these crystals determined what types of patterns would emerge in the spin system. Later, my work on the spontaneous localization of energy in driven-damped mechanical and electrical lattices again connected to the larger theme. I still remember my excitement when I finally realized that the localized patterns that spontaneously formed in my lattice could act back on the parts of the lattice (the individual nodes) to change their dynamics, and that this change would in turn cause the localized pattern be altered. What I was witnessing was multiple passes of the feedback loop playing out between two levels. The same resonances were also present in the work I did on synchronization of populations of phase oscillators. Individual oscillators following simple rules could collectively display modes of organization that were often quite surprising, and these modes were sensitive to the way in which oscillators had been coupled together. If the network structure was allowed to evolve in time and 'learn', then through the interplay of oscillator dynamics and network dynamics, we found that networks could actually *self-assemble* spontaneously.

So I can now appreciate that much of the research I have been involved in could be loosely summarized by the sentence 'the whole is more than the sum of its parts.' But, more importantly, this is increasingly the case for the work of my colleagues as well. In fact, some of the most exciting work in physics has this flavor. The fairly recent discovery of the Majorana fermion in atomically thin superconducting wire is one intriguing example [103]. The 2016 Nobel prize in physics was awarded for work on topologically interesting states of matter. Within a whole range of physics research fields, the recognition that emergence is a feature of our physical reality is shining through. Most of the exciting new problems involve complex networks, strongly interacting systems, nonlinearity, self-organization, and/or multiple scales. If 20th century physics was dominated by reductionism, it appears likely that this century will be the age of emergence.

Bibliography

1. R. Feynman, *The Feynman Lectures, Vol.1* (Addison-Wesley, Reading, 1963)
2. J. Horgan, *The End of Science: Facing the Limits of Knowledge in the Twilight of the Scientific Age* (Addison-Wesley, New York, 1997)
3. C. Koch, *Consciousness: Confessions of a Romantic Reductionist* (MIT Press, Cambridge, 2012)
4. P.W. Anderson, More is different. Science **177**, 393 (1972)
5. R. Laughlin, *A Different Universe* (Basic Books, New York, 2005)
6. C.L. Morgan, *Emergent Evolution* (William & Norgate, London, 1923), p. 23
7. R. Hoffmann, *The Same and Not the Same* (Columbia University Press, New York, 1995)
8. R.J. Campbell, M.H. Bickhard, Physicalism, emergence and downward causation. Axiomathes **21**, 33 (2011)
9. P.B. Anderson, C. Emmeche, N.O. Finnemann, P.V. Christiansen (eds.), *Downward Causation - Minds, Bodies and Matter* (Aahrus University Press, Aarhus, 2000)
10. P. Humphreys, How properties emerge. Philos. Sci. **64**, 1 (1997)
11. M. Bitbol, Ontology, matter and emergence. Phenomenol. Cogn. Sci. **6**, 293 (2007)
12. G.K. Gyatso, *Modern Buddhism* (Tharpa, London, 2011)
13. D. Hofstadter, *Gödel, Escher, Bach* (Vintage, New York, 1979)
14. J. Kim, *Mind in a Physical World* (MIT Press, Cambridge, 1998)
15. T. Nagel, *Mind and Cosmos: Why the Materialist Neo-Darwinian Conception of Nature Is Almost Certainly False* (Oxford University Press, Oxford, 2012)
16. J. Wahman, *Narrative Naturalism: An Alternative Framework for Philosophy of Mind* (Lexington Books, Lanham, 2015)
17. P.S. Laplace, *A Philosophical Essay on Probabilities*, translated into English from the original French 6th edn. by F.W. Truscott and F.L. Emory (Dover, New York, 1951), p. 4
18. J. Gleick, *Chaos: Making of a New Science* (Vintage, New York, 1987)

19. É. Durkheim, *The Elementary Forms of the Religious Life* (Free Press, New York, 1912), p. 486

20. R. Feynman, *QED - The Strange Theory of Light and Matter* (Princeton University Press, Princeton, 1985)

21. G. Gabrielse, D. Hanneke, T. Kinoshita, M. Nio, B. Odom, New determination of the fine structure constant from the electron g-value and QED. Phys. Rev. Lett. **97**, 030802 (2006)

22. B. Abbott et al., Observation of a kilogram-scale oscillator near its quantum ground state. N. J. Phys. **11**, 073032 (2009)

23. T. Moore, *Six Ideas that Shaped Physics, Unit Q* (McGraw-Hill, Boston, 2003)

24. C.A. Fuchs, N.D. Mermin, R. Schack, An introduction to QBism with an application to the locality of quantum mechanics. Am. J. Phys. **82**, 749 (2014)

25. B.J. Pearson, D.P. Jackson, A hand-on introduction to single photons and quantum mechanics for undrgraduates. Am. J. Phys. **78**, 1 (2010)

26. R. Feynman, *The Feynman Lectures, Vol.3* (Addison-Wesley, Reading, 1963), pp. 1–9

27. J. Bell, On the problem of hidden variables in quantum mechanics. Rev. Mod. Phys. **38**, 447–452 (1966)

28. A. Aspect, J. Dalibard, G. Roger, Experimental test of Bell's inequalities using time-varying analyzers. Phys. Rev. Lett. **49**, 1804 (1982)

29. N.D. Mermin, Bringing home the quantum world: quantum mysteries for anybody. Am. J. Phys. **49**, 940 (1981)

30. A. Einstein, B. Podolsky, N. Rosen, Can quantum-mechanical description of physical reality be considered complete? Phys. Rev. **47**, 777–780 (1935)

31. G. Greenstein, A.G. Zajonc, *The Quantum Challenge* (Jones and Bartlett, Sudbury, 1997)

32. N.D. Mermin, Is the moon there when no-one looks - reality and the quantum theory. Phys. Today **38**, 38 (1985)

33. E. Schrödinger, The present situation in quantum mechanics. Proc. Am. Philos. Soc. **124**, 323 (1935)

34. F.M. Kronz, J.T. Tiehen, Emergence and quantum mechanics. Philos. Sci. **69**, 324 (2002)

35. B. Greene, *The Fabric of the Cosmos*, Chap. 4 (Vintage, New York, 2007)

36. C. Cercignani, *Ludwig Boltzmann - The Man Who Trusted Atoms* (Oxford University Press, Oxford, 1998)

37. P.C.W. Davies, *The Physics of Time Asymmetry* (University of California Press, Berkeley, 1974)

38. J.J. Sakurai, *Modern Qauntum Mechanics*, Chap. 4 (Addison-Wesley, Reading, 1994)

39. George Johnson, *Strange Beauty*, Chap. 7 (Vintage, New York, 2000)

40. S. Brush, N. Hall, *The Kinetic Theory of Gases* (Imperial College Press, London, 2003), p. 262f [Reprinted]
41. I. Prigogine, *The End of Certainty* (The Free Press, New York, 1996)
42. C. Kittel, *Introduction to Solid State Physics*, 7th edn. (Wiley, New York, 1996), p. 548
43. L.Q. English, M. Sato, A.J. Sievers, Modulational instability of the nonlinear uniform spin-wave mode in easy axis antiferromagnetic chains. Phys. Rev. B **67**, 024403 (2003)
44. P.W. Anderson, *More and Different - Notes from a Thoughtful Curmudgeon* (World Scientific, Singapore, 2011)
45. M. Faraday, T. Northmore, On the liquefaction and solidification of bodies generally existing as gases. Philos. Trans. **135**, 155–177 (1845)
46. D. van Delft, Little cup of helium, big science. Phys. Today **61**, 36 (2008)
47. D. van Delft, P. Kes, The discovery of superconductivity. Phys. Today **63**, 36 (2010)
48. T. Hargittai, M. Hargittai, *Candid Science VI* (Imperial College Press, London, 2006), pp. 710–731
49. D.D. Osheroff, R.C. Richardson, D.M. Lee, Evidence for a new phase of solid He^3. Phys. Rev. Lett. **28**, 885 (1972)
50. A. Lesne, M. Laguës, *Scale Invariance - From Phase Transitions to Turbulence*, Chap. 3 (Springer, Heidelberg, 2003)
51. K.G. Wilson, Problems in physics with many scales of length. Sci. Am. **241**, 140 (1979)
52. D.V. Schroeder, *An Introduction to Thermal Physics* (Addison Wesley Longman, Harlow, 2000), p. 346ff
53. R.B. Laughlin, D. Pines, Proc. Natl. Acad. Sci. **97**, 28 (2000)
54. R.W. Battermann, Emergence, singularity and symmetry breaking. Found. Phys. **41**, 1031 (2011)
55. J.K. Jain, The composite fermion: a quantum particle and its quantum fluid. Phys. Today **53**, 39 (2000)
56. V.J. Goldman, B. Su, Resonant tunneling in the quantum hall regime: measurement of fractional charge. Science **267**, 5200 (1995)
57. R. de-Picciotto, M. Reznikov, M. Heiblum, V. Umansky, G. Bunin, D. Mahalu, Direct observation of fractional charge. Science **389**, 162 (1997)
58. G.P. Collins, Fractionally charged quasiparticles signal their presence with noise. Phys. Today **50**, 17 (1997)
59. H.L. Stormer, D.C. Tsui, A.C. Gossard, The fractional quantum Hall effect. Rev. Mod. Phys. **71**, S298 (1999)
60. P.W. Anderson, When the electron falls apart. Phys. Today **50**, 42 (1997)
61. J. E. Wesfreid, Scientific biography of Henri Bénard," in *Dynamics of Spatio-Temporal Cellular Structures*, vol. 207 (Springer, New York, 2006)

62. E.L. Koschmieder, Bénard convection. Adv. Chem. Phys. **26**, 177 (1974)
63. M.C. Cross, P.C. Hohenberg, Pattern formation outside of equilibrium. Rev. Mod. Phys. **65**, Chapter 8 (1993)
64. K.M.S. Bajaj, N. Mukolobwiez, N. Currier, G. Ahlers, Phys. Rev. Lett. **83**, 5282 (1999)
65. A. Juarrero, Top-down causation and autonomy in complex systems, in *Downward Causation and the Neurobiology of Free Will*, ed. by N. Murphy et al. (Springer, Berlin, 2009)
66. W.T. Newsome, Human freedom and 'emergence', in *Downward Causation and the Neurobiology of Free Will*, ed. by N. Murphy et al. (Springer, Berlin, 2009)
67. I. Stewart, Fearful symmetry. New Statesman, pp. 39–41, 25 April 2011
68. S. Strogatz, *Sync: The Emerging Science of Spontaneous Order* (Hyperion, New York, 2003)
69. S. Strogatz, TED Talk, February 2004
70. H. MacDonald, The human flock. The New York Times Magazine, 6 December 2015
71. E.R. Kandel, J.H. Schwartz, T.M. Jessel, *Principles of Neural Science*, 4th edn. (McGraw-Hill, New York, 2000)
72. F. Khalili-Araghi et al., Biophys. J. **98**, 2189–2198 (2010)
73. P. Dayan, L.F. Abbot, *Theoretical Neuroscience* (MIT Press, Cambridge, 2005)
74. E.M. Izhikevich, *Dynamical Systems in Neuroscience* (MIT Press, Cambridge, 2010)
75. D. Watts, *Six Degrees - The Science of a Connected Age* (W.W. Norton, New York, 2003)
76. T. Parsons, *The Social System* (The Free Press, New York, 1951)
77. D. Elder-Vaas, *The Causal Power of Social Structure* (Cambridge University Press, Cambridge, 2010)
78. The CIA World Factbook, North Korea (2013)
79. The Human Rights Watch, 1999 Report. http://www.hrw.org/reports/1999/rwanda/Geno1-3-09.htm
80. T. O'Neill, Escape from North Korea. National Geographic Magazine, February 2009 issue
81. In 1999, the BBC reported an estimate of 3 million casualties. http://news.bbc.co.uk/2/hi/asia-pacific/281132.stm
82. J. Diamond, *Guns, Germs and Steel* (W.W. Norton, New York, 1997)
83. D. Acemoglu, J. Robinson, *Why Nations Fail* (Crown, New York, 2012)
84. P. Connolly, A. Smith, B. Kelly, *Too Young to Notice? The Cultural and Political Awareness of 3–6 Year Olds in Northern Ireland* (Community Relations Council, Belfast, 2002). http://www.paulconnolly.net/publications/report_2002a.htm
85. P. Brownson, A. Marryman, See baby discriminate. Newsweek Magazine, September 2009

86. C. Steele, J. Aronson, Stereotype threat and the intellectual test performance of African Americans. J. Pers. Soc. Psychol. **69**, 797–811 (1995)
87. The Stone, Women in philosophy - Do the math; The disappearing women. The New York Times, September 2013
88. N. Ambady, M. Shih, A. Kim, T.L. Pittinsky, Stereotype susceptibility in children. Psychol. Sci. **12**, 385–390 (2001)
89. T. Kuhn, *The Structure of Scientific Revolutions* (University of Chicago Press, Chicago, 1962)
90. E. van den Haag, On faith, in *Religious Experience and Truth*, ed. by S. Hook (New York University Press, New York, 1961)
91. P. Tillich, *Dynamics of Faith* (Harper & Brother, New York, 1957)
92. N. deGrasse Tyson, Real Time with Bill Maher, October 2012
93. J. Polkinghorne, *Science and Religion in Quest of Truth* (Yale University Press, New Haven, 2011)
94. S. Weinberg, *Dreams of a Final Theory* (Random House, New York, 1994)
95. S. Kauffman, Beyond reductionism: reinventing the sacred (2006). Edge.com (Edge Foundation). Retrieved 24 Apr 17
96. C. Clinton, *Harriet Tubman* (Little, Brown, New York, 2004)
97. S. Bradford, *Harriet Tubman* (Corinth Books, New York, 1961)
98. J. Hick, *Faith and Knowledge* (Cornell University Press, Ithaca, 1966)
99. R. Abelson, The logic of faith and belief, in *Religious Experience and Truth*, ed. by S. Hook (New York University Press, New York, 1961)
100. J. Smith, *The Analogy of Experience: An Approach to Understanding Religious Truth*, Chap. I (Harper & Row, New York, 1973)
101. J. Hick, Meaning and truth in theology, in *Religious Experience and Truth*, ed. by S. Hook (New York University Press, New York, 1961)
102. G.K. Gyatso, *Joyful Path of Good Fortune*, 2nd edn. (Tharpa, London, 1995)
103. S. Nadj-Perge, I.K. Drozdov, J. Li, H. Chen, S. Jeon, J. Seo, A.H. MacDonald, B.A. Bernevig, A. Yazdani, Observation of Majorana fermions in ferromagnetic atomic chains on a superconductor. Science **346**, 602–607 (2014)

Index

© Springer International Publishing AG 2017
L.Q. English, *There Is No Theory of Everything*,
DOI 10.1007/978-3-319-59150-6

Printed in the United States
By Bookmasters